Fungal Biotechnology

Based on the proceedings of a meeting held in September 1978 and jointly sponsored by the British Mycological Society and the Society of Chemical Industry

THE BRITISH MYCOLOGICAL SOCIETY
SYMPOSIUM SERIES NO. 3

Fungal Biotechnology

edited by

J.E. SMITH
D.R. BERRY
B. KRISTIANSEN

Department of Applied Microbiology
University of Strathclyde, Glasgow

1980

ACADEMIC PRESS
A Subsidiary of Harcourt Brace Jovanovich, Publishers
London New York Toronto Sydney San Francisco

ACADEMIC PRESS INC. (LONDON) LTD.
24/28 Oval Road,
London NW1

United States Edition published by
ACADEMIC PRESS INC.
111 Fifth Avenue
New York, New York 10003

British Library Cataloguing in Publication Data
Fungal biotechnology. – *(British Mycological Society. Symposium series; no.3).*
1. Fungi – Industrial appliances – Congresses
I. Smith, John Edward II. Berry, David Richard III. Kristiansen, B IV. Society of Chemical Industry V. Series
602′.8 QR53 80-40585

ISBN 0-12-652950-7

Printed in Great Britain

CONTRIBUTORS

ANDERSON, J.G., Department of Applied Microbiology, University of Strathclyde, Glasgow G1 1XW.

ATKINSON, B., Department of Chemical Engineering, UMIST, P.O. Box 88, Manchester M60 1QD.

BALL, C., Glaxo Operations (UK) Limited, Ulverston, Cumbria, LA12 9DR.

BLAIN, J.A., Department of Biochemistry, University of Strathclyde, Glasgow G1 1XW.

BU'LOCK, J.D., Microbial Chemistry Laboratory, Department of Chemistry, University of Manchester, Manchester M13 9PL.

COCKER, R., Glaxo Operations (UK) Limited, Ulverston, Cumbria, LA12 9DR.

CORBETT, K., Beecham Pharmaceuticals, Clarendon Road, Worthing Sussex.

EKUNDAYO, J.A., Department of Biological Sciences, University of Lagos, Lagos, Nigeria.

EVELEIGH, D.E., Department of Biochemistry and Microbiology, Rutgers - The State University, New Brunswick, NJ 08903, USA.

GREENSHIELDS, R.N., Department of Biological Sciences, The University of Aston in Birmingham, Birmingham B4 7ET.

HAYES, W.A., Department of Biological Sciences, The University of Aston in Birmingham, Birmingham B4 7ET.

HOCKENHULL, D.J.D., Glaxo Operations (UK) Limited, Ulverston, Cumbria LA12 9DR.

JEFFRIES, T., United States Department of Agriculture, Forest Products Laboratory, P.O. Box 5130, Madison, WI 53705, USA.

KRISTIANSEN, B., Department of Applied Microbiology, University of Strathclyde, Glasgow G1 1XW.

LEWIS, P.J.S., Wander Foods Limited, Station Road, Kings Langley, Hertfordshire.

MICHALSKI, H., Lodz Technical University, 90-924 Lodz, Poland.

MONTENECOURT, B.S., Department of Biochemistry and Microbiology Rutgers - The State University, New Brunswick, NJ 08903, USA.

PACE, G.W., Biospecialities, Penrhyn Road, Knowsley Industrial Estate, Prescot, Merseyside L34 9HY.
SINCLAIR, C.G., Department of Chemical Engineering, UMIST, P.O. Box 88, Manchester M60 1QD.
SOLOMONS, G.L., RHM Research Limited, The Lord Rank Research Centre, Lincoln Road, High Wycombe, Bucks HP12 8QR.

PREFACE

The Symposium held in Glasgow in September 1978, under the auspices of the British Mycological Society and the Society for Chemical Industry, brought together academic and industrial workers representative of most of the main areas of fungal biotechnological research. The rapid expansion of biotechnology in recent years has lead to a greater awareness of the use of filamentous fungi in commercial processes. The morphology of these organisms presents many unique problems when they are grown in fermenters. In this volume we have assembled a series of chapters concerned with the influence of fungal growth on the construction and operation of liquid and solid-state fermenters. In particular, emphasis is placed on those processes which have found special application for growth and exploitation of filamentous fungi. The importance of the developments in fungal biotechnology throughout the world are indicated in chapters from contributors who have experience of biotechnological developments in different regions of the world.

We regret that some of the original conference papers are not included since the contributors were unable or unwilling to fulfil their previous agreement to prepare a written contribution. Replacing this material with new chapters has caused a regrettable delay in publication. However, we would like to thank all contributors for producing such stimulating and well prepared manuscripts.

We must also acknowledge our indebtedness to Mrs. Margaret Provan for typing the text and for her continued help during the organisation of the Symposium and the compilation of the volume. Our thanks also go to Miss Isobel Docherty for art work and to Dr. J.G. Anderson for assistance in certain aspects of editing.

John E. Smith
David R. Berry
Bjørn Kristiansen

May, 1980.

CONTENTS

INOCULUM DEVELOPMENT WITH PARTICULAR REFERENCE TO ASPERGILLUS AND PENICILLIUM

D.J.D. Hockenhull

Glaxo Operations (UK) Limited,
Ulverston, Cumbria LA12 9DR

INTRODUCTION

In the fermentation industry, current manufacturing practice demands the contact of large amounts of biomass with air and nutrient in huge reactors (*c* 50-100 m^3). The biomass has to be of the right kind and in the right state. The review which follows will describe the technology needed to ensure that these requirements are supplied in the best possible way, taking into account differences between organisms and the nature of the processes they are involved in. Biomass with highly efficient and predictable performance, and deriving from a very small quantity of very high quality seed, is what is wanted. This implies, as we shall see, that the preconditioning of the biomass through its various seed stages continues through to the production phase. The history of a culture may have so great an effect on its behaviour that subsequent environmental control is unable to evoke the desired pattern of performance. In a similar context, I have said according to Calam's record 'Once a fermentation has been started, it can be made worse but not better' (Calam, 1976). While I no longer think this to be strictly true, treating cultures as though it were can often save a lot of grief. Bearing in mind the present technical level of the production stage of most commercial fermentations, current practice in inoculum development comes as a shock. Empiricism reigns supreme. Even extensive reviews such as that of Meyrath & Suchanek (1972) give little indication of a more structured project. Media are generally hallowed by long usage or are translated from the technology of an older project, though they may be none the worse for that. Criteria of fitness for transfer to the next scale of propagation are mostly arbitrary. We need uncontaminated cultures as free from variants as is humanly possible. We need a technology for such cultures to adequate biomass and we need to ensure that, at each transfer, the correct conditions are enforced. The ensuing discussion deals with fulfilling these desiderata so that the final outcome - predictable excellence in the production stage - is guaranteed.

THE MASTER CULTURE AND CULTURAL STOCKS

The ideal stock should consist of similar individuals differing from each other in no respect but age. It should be free from contaminants and variants of the master organism. This, itself, should therefore be beyond suspicion. To ensure this as far as is practicably possible, one should isolate a single organism and, from this, establish a small master culture of sufficient size to permit progeny testing before using it to make working stock cultures or sub-masters. These master cultures must be available over a period of one or two years, although good practice would dictate that new master cultures be ready to replace the originals at rather shorter intervals. Long storage times demand special techniques and, for practical purposes, the establishing of sub-master cultures to provide tested working stocks.

In general, one is fortunate with industrial organisms, in that most of them form spores which have much more satisfactory storage properties than the vegetative thallus. Detailed procedures have been outlined for producing proven inocula from unproven culture material. Methods of storage, and the criteria by which they should be judged will also be described in this article. In what is probably the most important contribution to this topic, Lincoln (1960) sets a standard of acceptable purity as being zero external contamination and 5% colonial or observable genetic variation. The latter would appear to be a high percentage for most fungal processes but the user must himself eventually set standards appropriate to the growth stage and the overall biomass increases in his process.

PRESERVATION OF STOCK CULTURES

A long shelf life, good germination and stable progeny are the hallmarks of a workable storage system. The subject has been reviewed in detail by Lincoln (1960); Fennell (1960); Muggleton (1963); Calam (1964); Casida (1968); Onions (1971) and Hesseltine & Hayes (1973).

Fungi are best stored as conidiospores because of their resistance to adverse conditions such as cold, low water activity or lack of nutrients. The preservation techniques in use today are (1) lyophilization, (2) drying in soil and (3) freezing with liquid nitrogen. These techniques are described in some detail by Muggleton (1963).

In lyophilization (freeze-drying) the spores are suspended in a carrier, or protective agent, such as bovine serum or skimmed milk, and filled into ampoules. They are rapidly frozen to low temperature (-40°C) and dried under vacuum. The ampoules are then sealed and stored at 4-10°C. The spores should remain viable for long periods. Lincoln (1960) has stated that storage loss should not exceed 5% per annum. When needed, the cultures can be revived by re-suspension in growth medium.

Soil flasks require a chosen rich garden soil which is dried and sieved to a suitable particle size, adding fine sand and/or chalk if necessary. The techniques employed have been described by Greene & Fred (1934) and by Onions (1971). The prepared soil is distributed into tubes or flasks and autoclaved several times to ensure absolute sterility. A small volume of a spore suspension is then added and the moisture removed by evacuation over a desiccant. Dried and sealed, the soil flasks or tubes are stored at room temperature. Viability of fungal spores is generally good.

Both these methods are becoming less popular than storage in or over liquid nitrogen (-165 to -195°C). The techniques have been described by Macdonald (1968, 1972) for *Penicillium chrysogenum*. Sealed ampoules of spores suspended in a suitable medium (e.g. 10% glycerol) are quickly frozen in nitrogen and stored therein until needed. For safety, storage over liquid nitrogen rather than in it is preferable. Cracked ampoules stored beneath the surface can fill up with liquid nitrogen. Removal to a warm atmosphere rapidly converts the ampoule into a highly effective bomb. Cultures are resuscitated by allowing the ampoule to warm up to room temperature before unsealing and adding the contents to fresh medium. Macdonald (1972) claimed a high viability and low mutation rate when he stored spores of *P. chrysogenum* at -196°C. If the culture to be preserved does not form adequate numbers of spores, vegetative cells are preserved. Although by no means so suitable as spores, hyphal fragments can be similarly stored. The liquid nitrogen technique has been found most practical. Not least, it avoids the need for serial transfer of vegetative material as a means of keeping a culture alive, which calls for many more checks on identity and behaviour and for frequent re-isolation. In some instances, however, those methods are still necessary.

GERMINATION OF SPORES

The development of spores from vegetative mycelium is accompanied by a fall in biochemical activity. Smith (1975) states "all fungal spores are considered to show some degree of dormancy since they are no longer engaged in the synthesis of new cellular material". Respiration is slow and the enzyme concentration and activity much reduced. For the spore to give rise to a rapidly growing culture, these processes need restoration. That takes a long time compared with normal vegetative reproduction. The lag involved is called the germination time, the recovery from spore to active vegetative growth being likened to germination (Fig. 1 and 2).

In addition to this basic phenomenon, many spores contain inhibiting substances that prevent them from responding prematurely to promising but dubious or transitory conditions. One of these has been described for the rust fungus, *Puccinia graminis* (Forsyth, 1955). The rate of removal of the inhibitor (which was shown to be trimethylethylene) depended on the number of spores, the degree and kind of washing, the pH value of the suspension and other conditions. Complete removal of the inhibitor allowed germination to occur quickly and completely. Scott, Alderson & Papworth (1972) observed similar phenomena in the conidia of *Aspergillus nidulans*.

A complementary function was demonstrated in *Aspergillus amstelodami* by Darling & McArdle (1959). They described the production of a substance from mycelial growth which stimulated or overcame the reluctance of spores to germinate. Vegetative growth was unaffected.

A considerable amount of research has been carried out in this field but little has resulted in the way of application. Media are still chosen pragmatically. However, the empirical formulations, especially of complex media, may well be the most useful in supplementing depleted intermediate pool levels. The field of spore germination in *Aspergillus* has recently been reviewed by Smith & Anderson (1973).

VEGETATIVE PROPAGATION

There are two aims in inoculum development: that of producing the right type of inoculum for the final stage of production, and that of getting enough of it. The latter is a necessary concomitant of the increase in scale incumbent on modern fermentations.

The simplest scheme uses vegetative seed stages on similar media right through from the storage ampoule to the production unit (Fig. 3, 4 and 5). The aim is vigorous, continuous growth right through each stage at very nearly uniform rate

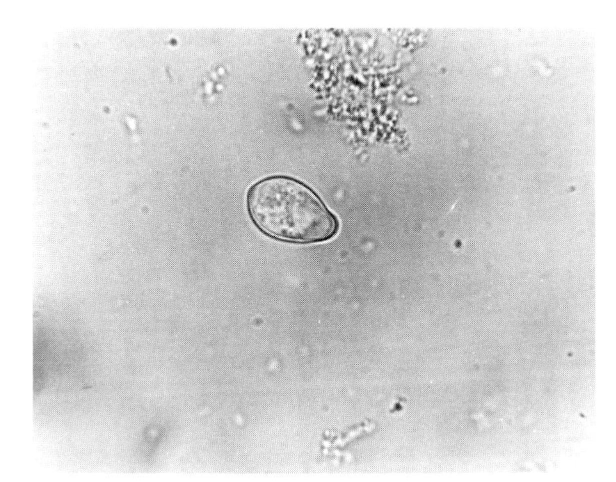

Fig. 1. *Penicillium chrysogenum* germinating spore x 2080.

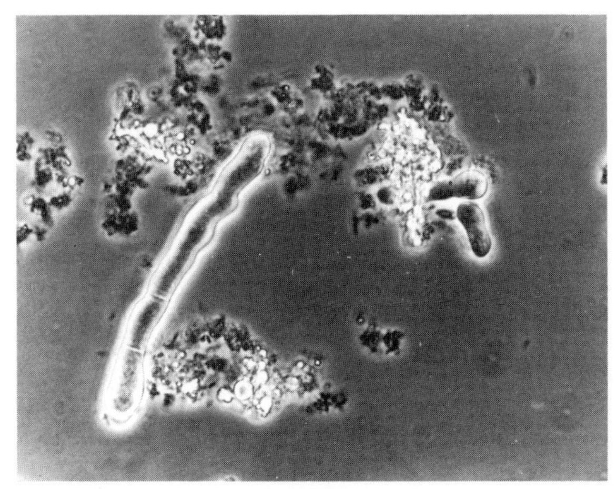

Fig. 2. *Penicillium chrysogenum* spores with germ tubes x 1088 (phase contrast)

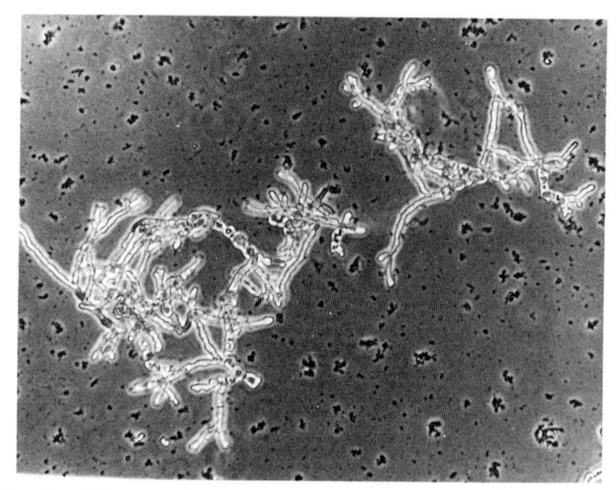

Fig. 3. *Penicillium chrysogenum* vegetative growth: inoculum state. 48 hours x 272 (phase contrast).

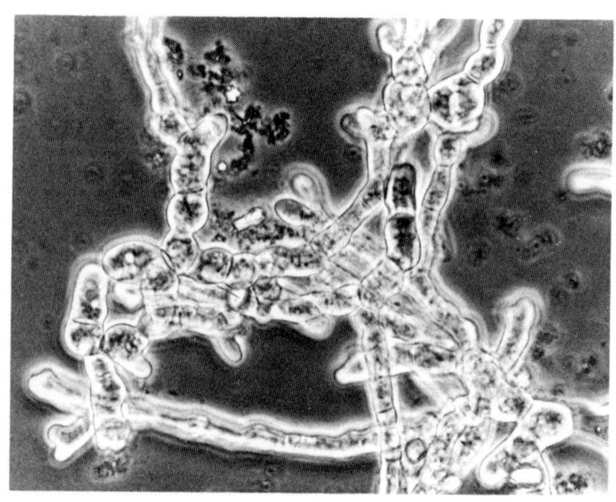

Fig. 4. *Penicillium chrysogenum* as Fig. 3 x 1088 (phase contrast.

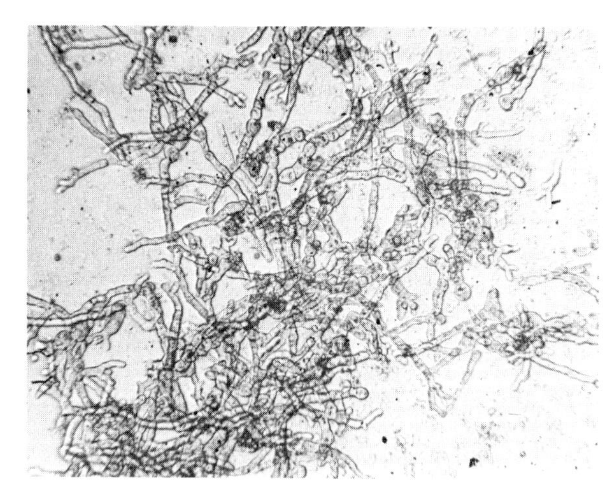

Fig. 5. *Penicillium chrysogenum* as Fig. 3 x 480 (bright field).

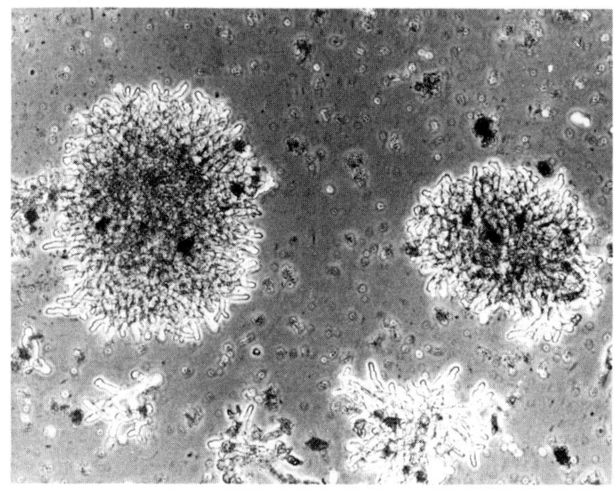

Fig. 6. *Penicillium chrysogenum* pellet growth x 272 (phase contrast).

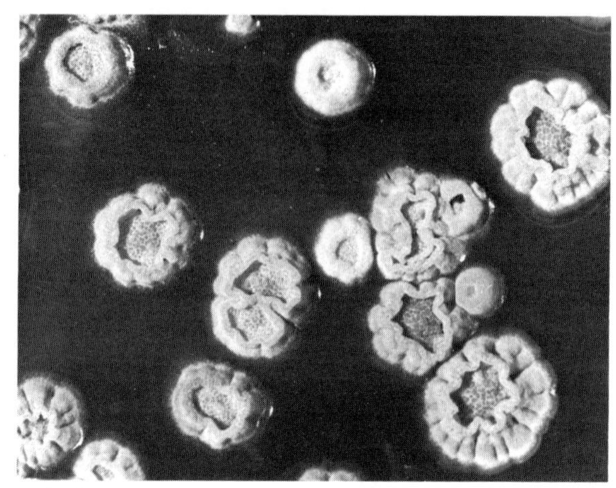

Fig. 7. *Penicillium chrysogenum* colonies of mixed types x 2.8

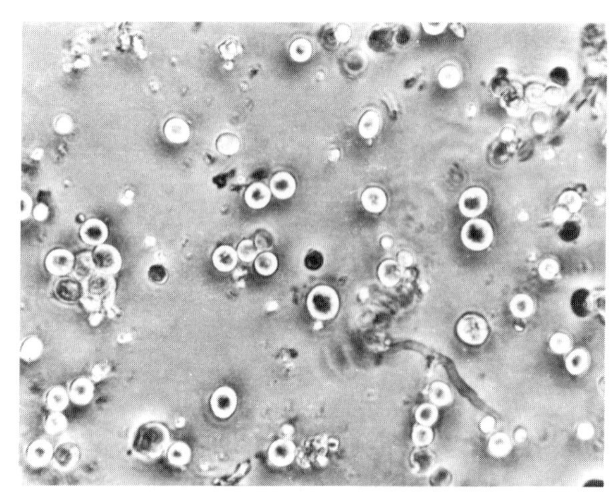

Fig. 8. *Penicillium chrysogenum* Submerged spores x 800 (phase contrast).

$(\frac{1}{x}\frac{dx}{dt} \simeq \mu_{max})$. The medium should not become exhausted or highly unbalanced (low pH, etc.) before transfer at any stage. In most practical instances, the desirable amount of inoculum rate is at least 5% of the next stage volume (10% being preferred). Therefore, from the first spore suspension to the final unit in which the desired product is made, the growth could involve passage through a series of six or seven increasing stages. One particular disadvantage of this lies in the use, e.g. as propagators, of intermediary 5 to 10 litre vessels. They consume a great deal of labour and involve special sterility hazards both when inoculating them and when transferring them to the next stage, especially as they are not mobile or near to production units and thus may require transport in intermediary containers.

A rough and ready way of avoiding this is to omit these intermediate stages altogether. The inoculum rate may thereby be reduced to the order of 1%. Such a practice is open to criticism on the grounds that there may be more selection pressure at such low levels of inoculum. Moreover, the larger propagator is, of course, in use for a longer period. On the other hand, handling costs are lower. When propagation is carried out substantially in the vegetative state, the choice of medium is important. Hesseltine & Haynes (1973) say "for the maintenance of stock cultures a chemically undefined, but reproducible, stock medium is better than a synthetic one". Their reasoning is that an organism growing in nature almost invariably does so on an undefined substrate, and that a defined medium is more prone to select variants. They also observe that a stock medium should be no 'richer' than is necessary to perpetuate the culture without change. Thus glucose is not used, or is used only to a level at which its drastic lowering of the pH value can be controlled by buffering. Alternatively, a complex medium can contain levels of carbon and nitrogen sufficient to support maximum growth until transfer, as well as a mixture of acidogenic and alkaligenic ingredients such as to permit of substantial growth without much pH drift. The medium should, however, contain sufficient carbon for the organism to stay in the vegetative growth stage, and for the catabolic repression of secondary enzyme systems. If the organism grows so slowly that it permits a variety of such enzymes, selection can operate on the differences between organisms and exaggerate the heterogenity of the culture.

An illuminating series of experiments on *P. chrysogenum* were carried out by Righelato (1976). He showed that continuous culture of the organism under carbohydrate-limited conditions led to a loss of penicillin-synthesising power and an increase in the proportion of non-conidiated variants. This did not occur when the growth rate was limited by ammonia, phosphate or sulphate.

In many instances, the propagation medium is a simplified version of the production medium and often is modified in the ways suggested by Haynes & Hesseltine (1973). Let me instance an example from around 20 years ago. The composition of a production medium for penicillin was lactose 4.0%, CSL solids 3.0%, glucose 0.5%, phenylacetyl ethanolamide 0.2%, potassium dihydrogen phosphate 0.5%, precipitated chalk 0.5% and soya bean oil 0.25%; pH adjusted to 4.7 - 4.9. The corresponding seed medium was CSL solids 3.0%, sucrose 4.0%, chalk 0.75% and soya bean oil 0.75%; pH adjusted to 5.3 - 5.6. With relatively unimportant differences these media are still in use today.

The use of vegetative seed stages throughout has an advantage that the growth rate of the biomass continues steadily at something approaching its maximum, with negligible interruption on transfer from one size of vessel to another. Hence the total lapsed time is minimal and so are the number of generations. While these are considerable, taking approximately 20 generations (in 10 ml of suspension from the original ampoule) to serve a 5 m^3 seed vessel, the genetic selection pressure is less than with any other method. In practice, the inoculum for the final production culture is remarkably free from variants as we find it. This certainly shows that, under conditions of high carbon feed, the selection of low yielding mutants is minimal. On the other hand, the carbon-limited production stage gives rise to variants much more readily.

Not all cultures can maintain their growth rate between propagation stages. An important cause of this is pelleting (Fig. 6). The aggregation of biomass, whether loose or sclerotial, will result in fewer growth centres at transfer; also pellets take up oxygen and nutrient more slowly than does diffuse growth. In many instances, treating the culture in a Waring blender or similar device before transfer has been found advantageous. Savage & van der Brook (1946) homogenised pellet growth of *P. notatum* and *P. chrysogenum* from shake flask cultures. After two minutes of blending they found the increase in viable growth centres was such that an inoculum 50,000 times less than that of the unblended cultures gave equally good results. I have so far failed to find any reference to the blending of mycelial cultures; work in this area has perhaps been looked on unfavourably because of the added complication, sterility risk and mechanical destruction of the organism in the blending operation.

Reverting to the general topic of vegetative inoculum,

there is seldom time to detect contamination in the stages proximal to the production stage. Certainly it is true that one can hold up the middle stages - if the bulk is not too great - by chilling, but the effects on the culture has never been thoroughly evaluated for any organism and there may be dangers in inducing or selecting mutations.

In addition, if the vessel schedule has been upset by the loss of infected tanks, necessitating inoculum as an extra batch before or after the scheduled time for transfer, there is no flexibility built into that system. This is what leads us to a consideration of 'storage' steps as alternatives to some of the vegetative stages.

SPORULATION STEPS AS ALTERNATIVES TO VEGETATIVE PROPAGATION

Spores are admirably suited to interrupted culture sequences. They are resistant to adverse conditions, may be cold-stored dry or in suspension and give adequate viability compared with vegetative cultures. This indicates the possibility of holding back the main stream of inoculum preparation until samples have been tested for contamination or variants. In addition, at the master culture level, the preparation of a sporing sub-culture is a useful way of ensuring a working ampoule stock of submaster cultures. (Be warned, however, that some low-yielding variants are very heavy sporers; conditions favouring their production should be avoided). A further advantage of a sporulation step is that the number of growing points in a seed culture vessel can be much greater than in vegetative cells. This allows one to miss out on a stage or even two and advance from, at the greatest stretch, 500 ml to 500 litres without an intermediate vessel. In the example cited, the 5 and 50 litre stages, which are the most difficult to maintain sterile, the most finicky to run and the most exacting of labour and expense, may be cut out by the use of an appropriate preparation (whether prepared in solid medium or in deep culture). The additional value of such a step in programming is easy to see. Material from the ampoule is frequently inoculated into flat bottles of solid medium designed to give good conidiophore production. This should show uniform growth in which variation can be seen with tolerable ease (and may be confirmed by plating out an appropriately diluted suspension).

Table 1 represents typical solid state sporulation media. The penicillin medium is a simplified version, described by Booth (1971), of that of Moyer & Coghill (1946). A more recent medium based on tomato juice, peptone and molasses has been described by Segal & Johnson (1963). The medium for *Aspergillus niger* was recommended by Steel, Lenz & Martin (1954).

Table 1. Media for production of spores in agar gel

A. *Penicillium chrysogenum* (Booth, 1971; Moyer & Coghill, 1946)

	g/l
Glycerol	7.5
Cane molasses	7.5
Curbay BG	2.5
$MgSO_4\ 7H_2O$	0.05
KH_2PO_4	0.06
Peptone	5.0
NaCl	4.0
Agar	20

B. *Aspergillus niger* (Steel, Lenz & Martin, 1958)

	g/l
Molasses	300
KH_2PO_4	0.5
Agar	20

Both the media in Table 1 have a high salt content. The spores, thus grown, are prepared as a suspension for transfer to the next growth stage. This is done by washing them off the agar with water containing a non-toxic wetting agent (such as Tween 80 or Turkey Red Oil). Glass beads of 5 mm diameter, or similarly sized gravel, may be used to detach the conidiospores. A suspension made in this way is convenient to handle. It is also useful because tests for uniformity and sterility (Fig. 7) can be carried out as well as the determination of direct and viable counts.

The use of flat bottles for spore production has limitations. A single unit is hardly big enough to inoculate large volumes of broth, of the order of 500 l. True, several bottles may be bulked but more prolific methods are called for. As an example, let us consider a representative strain of *P. chrysogenum* (for various train lines show different abilities to sporulate). It could produce about 10^{10} spores on a 300 cm^2 agar layer in a Roux bottle.

PRODUCTION OF SPORES IN SUBMERGED CULTURE

Producing spores in submerged culture is convenient and elegant. In deep culture on appropriate medium, the *P. chrysogenum* culture we referred to previously will produce 10^{10} spores per litre of culture (Fig. 8). The technique was first described by Foster, McDaniel, Woodruff & Stokes (1945) with the closely related *P. notatum*. The production of submerged spores was induced by including calcium chloride (2.5%) in a defined nitrate-sucrose medium. The spore counts were as high as 4 x 10^8 spores per ml. Gilbert & Hickey (1946) obtained copious sporulation with a cornsteep liquor-lactose medium. The phenomenon has since been studied by Meyers & Knight (1958, 1961) in *P. roquefortii*; by Morton (1961) in *P. griseofulvium*; by Vézina, Singh & Sehgal (1965) and Vézina & Singh (1974) in *Aspergillus ochraceus*, *Fusarium moniliforme* and *Stachydium theobromae* and by Galbraith & Smith (1969) in *A. niger*. Spore production by submerged culture has great advantages in simplicity of operation, easier sterile transfer and the potential of larger scale operation. A further advantage to the technique, however, is that the spore suspension may be counted, so that inoculations are much more reproducible in character. Furthermore, by cold storage for a suitable period, the vegetative material may be reduced to give a highly uniform inoculum consisting mainly of spores. A price one may have to pay is that the technique does not produce as many spores as does surface growth, though this may reflect the 'state of the art' rather than the potential of the method. The conditions needed to produce a good crop of conidiospores in surface culture as well as in deep culture have been given considerable study. Smith & Anderson (1973) and Smith (1975) indicate that ammonia run-out is the primary event leading to conidiation in *Aspergillus*. That applies both to aerial conidia in surface growth and to conidia produced in deep culture. (Surface conidiation is helped, however, by the fact that the nitrogenous medium is filtered through the mycelial mat, and reduced thereby, before it reaches the aerial part of the thallus). In addition to this main cause, certain organic acids, glyoxylate in particular, but also some members of the tricarboxylic acid cycle, have been shown seemingly to neutralise the effect of ammonia and to encourage conidiation in deep culture. Galbraith & Smith (1969) did, in fact, induce sporulation in deep culture by the addition of oxoglutarate. A further factor seems to be the inclusion of salts, particularly chlorides, into the media. Table 2 gives a medium, found by Foster *et al.* (1945), to produce submerged spores of *P. notatum* and one used by Vézina *et al.* (1965) for the same purpose with *A. ochraceus*:

Table 2. Media for production of submerged spores

A. *Penicillium notatum* (Foster *et al.*, 1945)

Source	g/l
Sucrose	20
$NaNO_3$	6
KH_2PO_4	1.5
$MgSO_4\ 7H_2O$	0.5
$CaCl_2$	25

B. *Aspergillus ochraceus* (Vézina *et al.*, 1965)

Source	g/l
Glucose	25
NaCl	25
CSL	5
Molasses (Blackstrap)	50

You will note that high chloride levels are used in both. A point made by Vézina *et al.* (1968) is that "all fungi do not respond identically to external factors and great discrepancy is observed between species of the same genus and even between strains of the same species."

THE PRODUCTION-INOCULUM INTERFACE

This section will refer mainly to the behaviour of the production stage under the influence of the inoculum. Most of the fermentations we are concerned with are carried out for secondary product formation. The culture goes through two phases - distinct in some organisms, overlapping in others - which correspond first to growth and setting up the cultures in the correct biological condition; second, to the maintenance of a biomass, as large as can be adequately supplied with oxygen, in the correct nutritional state for rapid and prolonged production of the special product.

One can readily see then, that time can be saved on the first stage by using more inoculum. However, this leaves less time to direct the final culture into the optimal for product formation. The sophistication of the process has a profound effect on what is or is not possible and, by and large, the batch process, when the total ingredients are added before inoculation, does not do so well on such high inoculum rates as processes involving continuous feeding. Even in these instances, the conditioning or imprinting of metabolic pattern takes place powerfully from the earliest hours of growth. Especially as one approaches secondary product formation, when the specific growth rate might well have fallen to 10% of normal, does one find the special enzymes required for product formation more and more difficult to induce to the maximum. In the finality, the production process is a compromise between what is possible and what is desirable; small wonder then that fermentations for different products have solved their problems in different ways, mostly by trial and error.

As has been mentioned, the main intention has been to approximate the single stages of vegetative growth to a continuously ongoing culture at near to maximal growth rate. There are many exceptions to this aim. Allowing a culture to 'mature' although we cannot define this process - before transfer is often better. Meyrath & Suchanek (1972) (Fig. 9) show that the mycelial weight put on by a culture needed with an autolysing inoculum of *Aspergillus oryzae* is higher throughout than a similar culture seeded from similar inoculum (with the same content of dry matter) in the 'logarithmic' phase of growth. This kind of phenomenon is often encountered.

A further example of the imposition of a growth pattern that persists well into the next stage is the effect of inoculum size on the pH course of the culture. The activity of the enzymes releasing ammonia from the substrate increases faster, relative to the growth rate, than the rate of uptake of nitrogen for cell synthesis which is proportional to growth rate. Therefore, the culture with less active biomass produces more ammonia than it can use in growth. The pH value increases and can persist higher for at least a week (Fig. 10). This is similar in a way to the effect of pH on, for example, a *P. chrysogenum* culture inoculated at pH 4.7 as compared with one inoculated at 6.0. We will find a much greater stimulation of ammonia-releasing enzymes at the lower values, causing the pH to rise above that of the higher starter and persist at a higher value to the end of the exercise (Fig. 11). Recapitulating, the use of large inocula saves time but it leaves less scope for adaptation. Today we compensate for this by environmental control, that is, by a patterned feeding of desired ingredients.

TRANSFER CRITERIA

The most commonly used criterion for vegetative inocula is the biomass. Ideally, one would measure the dry matter

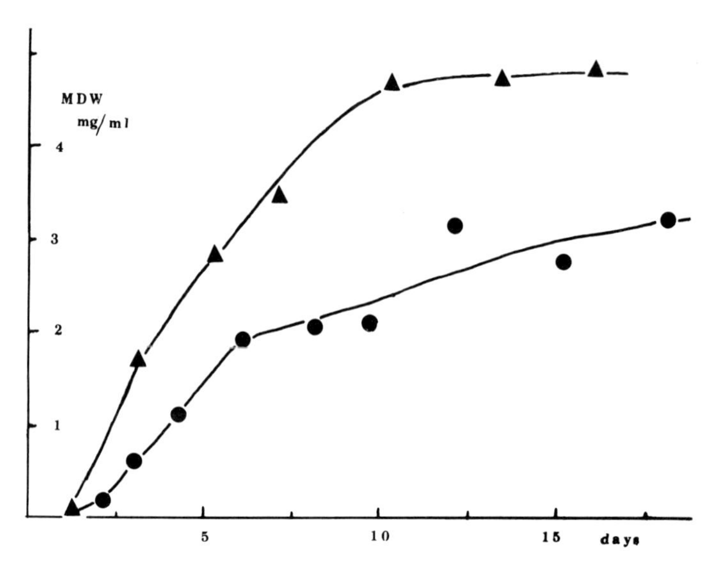

Fig. 9. Influence of the history of inoculum (equal size) on growth in advanced stages of culture development of *Aspergillus oryzae*. MDW = mycelial dry weight. ●—● , inoculum from growing phase; ▲—▲ , inoculum from autolytic phase. (Meyrath & Suchanek, 1972).

---

and transfer at an appropriate stage of growth. This is inconvenient because the determination takes a considerable time and also because many of the media employed contain particulate matter which interferes. Measurement of turbidity in fungi is not very easily or consistently determined because the mycelium settled quickly and homogeneous samples are difficult to take. Usually, the packed cell volume is the most useful criterion. It is usually determined by centrifuging under a set of standard conditions. If, however, the inoculum needs any maturing, transfer may have to be made long after the growth curve has levelled out. Then, the transfer is made at a fixed period after the culture has grown up to a certain level, or even by dead reckoning of growth time. Such calculations are supplemented by observation of pH, dO_2, CO_2 and by microscopical observations that may also be used as independent criteria.

Preparation of spores, whether as a suspension of aerial spores or as a submerged culture, are advantageous in a further respect; they can be held for a considerable time without appreciable alteration of subsequent performance.

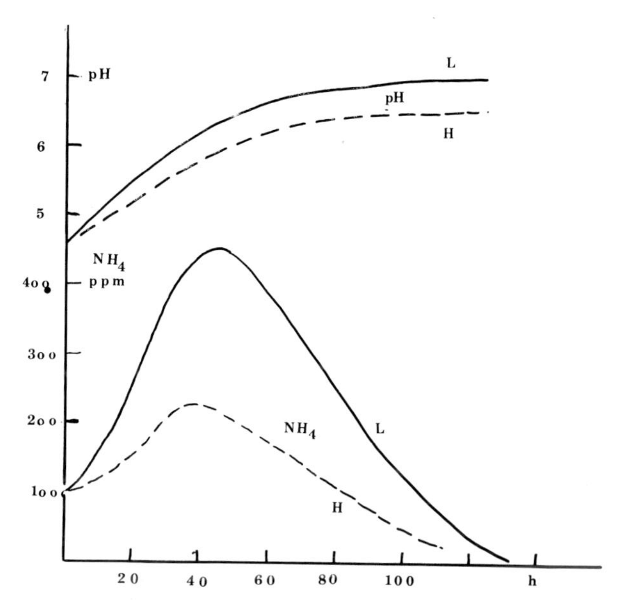

Fig. 10. Effect of inoculum size on subsequent pH behaviour. (*P. chrysogenum*). L, low initial pH value: H, high initial pH value.

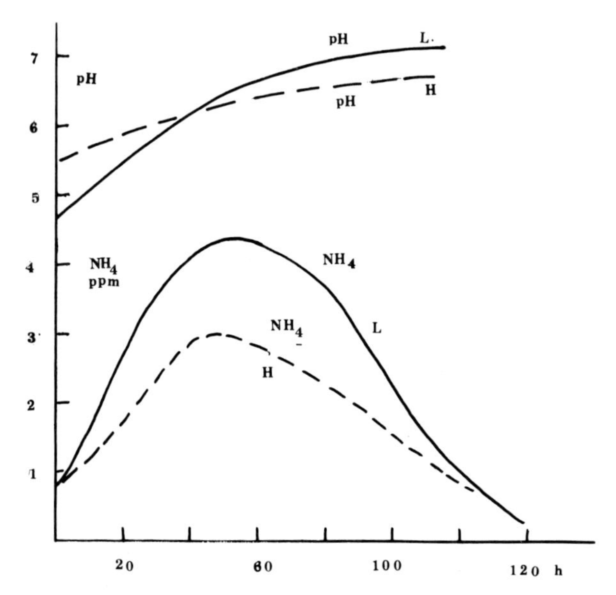

Fig. 11. Anomalous effect of low initial pH on subsequent pH behaviour *(P. chrysogenum)*. L, low initial pH value, H, high initial pH value.

PELLET GROWTH

The phenomenon of pellet growth in a wide variety of fungi has been well reviewed by Whitaker & Long (1973). Low inoculum levels, toxicity of the medium, pH and excess aeration were each discussed, together with the advantages or disadvantages of this mode of growth in respect of production performance. A paper by Metz & Kossen (1977) extends this with mathematical models of pelleted systems. These workers also discuss media composition, nitrogen metabolism, polymer additions, surface active agents and growth rate as further contributory factors.

The most common precipitating factor is too low an inoculum (Foster, 1939). This may occur with either vegetative or spore inocula. The latter, however, is more regular in its behaviour and the example quoted for *P. chrysogenum* by Calam (1976) implies this (Table 3). Similar results were obtained with *A. oryzae*. In *A. niger*, Steel, Lenz & Martin (1954) deliberately used low levels of spores to obtain pelleted production cultures, considered by the authors to produce citric acid more efficiently. Similar morphological patterns of the system had been obtained by Karow & Waksman (1947) with *A. wentii*.

Table 3. Effect of inoculum size on morphology and secondary product formation by *Penicillium chrysogenum* (Calam, 1976).

Concentration of spores units ml^{-1}	Morphology	Penicillin yield $\mu\ ml^{-1}$
10^2	dense pellets	500
10^3	" "	1800
2×10^3	open pellets	4000
10^4	confluent (filamentous)	5000

The pelleted form of morphology is more common in the early life of a culture; well-grown older cultures tend to be filamentous and are almost wholly so if the inoculum rate exceeds 10^6 spores per ml.

The advantages of the pellet form are that it permits a higher oxygen solution rate than the comparatively more viscous mycelial form. It must, however, pay for this by slower diffusion of air into the mould tissue. Which is the more important for any particular process can only be found out by experiment.

SUBSTRATE INCOMPATABILITY

If the media at any transfer stage differ markedly from each other in pH value, osmotic pressure or, especially in relative ionic strength or in using different anions, the viability of the biomass can be drastically altered. Ionic differences between the inoculum and the new medium set up electrical stresses across the cell wall because of the different transfer rates for different ions. In *P. chrysogenum* the sporulation stages introduce a special risk. Most of them contain a very high level of chloride and low phosphate, whereas the opposite is true for the production and vegetative stages. An experiment that will illustrate this was carried out by my colleague Dr. O'Callaghan (1977). When submerged spores, grown as is usual on a medium containing a high chloride level, were transferred to a normal type penicillin vegetative inoculum development medium based on cornsteep liquor and containing a very high level of phosphate as the predominate anion, there was immediate cell wall damage with the result that 90% of the culture failed to germinate.

In summary, therefore, the preparation and presentation of the inoculum, and its compatability with the production stage conditions, have profound effects on behaviour and yield. Incorrect handling may so badly affect the phenotype as to make it unresponsive to the environmental control being applied. So what Calam credits me with saying may be regarded as a good working rule - "once a fermentation has been started, it can be made worse but not better!"

APPENDIX

STERILITY AND ASEPSIS

There are relatively few commercial processes that will tolerate contamination. Inman (1971) has listed the possible types of disorder introduced by unwanted organisms:

1. Competition for nutrients, whether the contaminant overgrows the host or not.

2. Production of unfavourable conditions (*e.g.* low pH value, antibiotic formation, anaerobiosis).

3. Production of unwanted products, (*e.g.* pigments, foaming, etc.; modification of wanted products.)

4. Destruction of wanted product (*e.g.* contamination of penicillin producing cultures by penicillinase-producing organisms).

5. Reduction in extractability of product (*e.g.* viscous by-products or small organisms will affect broth filtration, or bring about emulsification of solvent, or engender difficulties in ion-exchange columns).

6. Public health hazard (undesirable organisms, unwanted antibiotic, pyrogen formation).

Even in non-sterile processes such as brewing and some single cell protein production, contaminant-free inoculum is mandatory. Special techniques like phosphoric acid washing are common in brewing for cleaning up the recycled yeast biomass. When the antibiotic industry changed over from surface to submerged culture, new problems of sterility were created which increased as the final stage fermenters got bigger.

Surface culture had been carried out usually in special mould culture flasks or quart milk bottles which had the advantage that they could be filled, inoculated, emptied and washed mechanically. The medium was inoculated with a spore suspension, some of which sank and some of which floated, to germinate and cover the surface with a ramifying growth that consolidated into a thick felt. Comparatively few spores (10^4 - 10^5 ml^{-1}) were needed for adequate growth. The spore suspension could tolerate 100 foreign organisms ml^{-1} and end up with 99% of the bottles sterile. (In actual fact, the rate of loss from infection was negligible even under the foulest conditions, as when inoculating in an open factory yard).

This contrasts greatly with the problems of keeping contaminants out of a 50 m^3 production unit with all its subordinate seed stages. For *P. chrysogenum* that size of tank demands 5 m^3 of inoculum containing 100 - 200 Kg of dry matter - an enormously greater amount (10^6 times, at least, one would guess.) The chances of contamination are consequently higher. The standards of hygiene, therefore, must be correspondingly improved. Not surprisingly, the establishment of an adequate inoculum service was one of the critical needs for the swift adoption of deep culture methods. It was just as well, at that stage, that we did not know the severity of the problem. In parenthesis, we probably get better sterility than we deserve because a single organism does not always generate a centre of infection.

As a general rule, bacteriologists are used to talking

about a minimum infective dose which, roughly speaking, means the lowest inoculum size that will 'take'. In considering the isolation of single organisms from the soil, Johnson (1972) discusses factors influencing the growth of small inocula. If one uses a synthetic medium the toxicity of trace metals may be too much for a small inoculum whereas a large one may either share out the toxic material or provide sufficient organic matter to complex with it. Often, too, organisms may need CO_2 at a critical concentration. With too few organisms and too high an aeration rate the culture may lose more CO_2 than it can generate, so that it dries off. Again, the media used for antibiotic culture have high salt contents or osmotic pressure or low pH values, or they contain antiseptics such as lactic acid. Or again, the organism may produce enough antibiotic to suppress many contaminants. However lucky we are in these respects, any plant will, after a time, build up around it foci of organisms that have similar growth characteristics to the production strains. These can be wild organisms of different types or poor yielding variants of the production cultures. The solution is the same - to develop adequate techniques of sterile handling.

Starting with uncontaminated master and submaster cultures, we must transfer and grow up the ensuing stages under ideal conditions. The basic essential is a clean, custom-built suite. The rooms should be provided with the minimum of the simplest furniture (benches, stools, etc.). Walls, ceilings and floors should be free from cracks, nicks and discontinuities that might harbour unwanted organisms. The rooms should be designed to be easily cleaned (Davis, 1968) and should be cleaned at regular intervals. There should be a steady flow of sterile air at all times. In addition, concinuous exposure of the area to ultra-violet radiation when not in use further cuts down the infective count. The subject has been well reviewed and illustrated by Austin &Timmerman (1965) and Austin (1967).

Nowadays, a great deal of culture manipulation is carried out in laminar-flow cabinets which provide a further degree of security. That subject has been well reviewed by McDade, Phillips, Sivinski & Whitfield (1969).

None of these precautions is any good if the staff who operate them are not adequately trained and equipped. They should be provided with clean overalls, gloves, caps and masks to prevent contamination brought in on clothing or by droplet infection from coughs and sneezes. Quick and deliberate movement should be the order of the day: 'the tempestuous petticoat' disturbs not only Herrick but also the settled organisms in the dust.

To maintain high standards, the air needs to be regularly

tested for infective organisms. Exposure of suitable agar plates has been improved on by the use of more modern equipment such as the slit sampler. Swabs of various surfaces may also be taken at appropriate intervals. Lastly, a necessary precaution to avoid cross-contamination between one's own seed stocks is to make only one organism available for handling at a time.

ACKNOWLEDGEMENT

My thanks are due to my colleagues, Dr. C. Ball, Mr. R.W. Larner and Dr. P.J. Bailey for help and discussion, and to Mr. H.J. Whalley for the photomicrographs.

REFERENCES

AUSTIN, P.R. (1967). *Clean rooms of the world. A casebook of 200 clean rooms*. Ann Arbor: Ann Arbor Publishing Company.

AUSTIN, P.R. & TIMMERMAN, S.U. (1965). *Design and operation of clean rooms*. Detroit: Business News Publishing Company.

BOOTH, C. (1971). Fungal culture media. *Methods of Microbiology* Volume 4, pp 49-94. Edited by C. Booth. London: Academic Press.

CALAM, C.T. (1964). The selection, improvement and preservation of microorganisms. *Progress in Industrial Microbiology* 5, 1-53.

CALAM, C.T. (1976). Starting investigational and production cultures. *Process Biochemistry* (April) pp 7-12.

CASIDA, L.E. (1968). *Industrial Microbiology*. New York: John Wiley.

DARLING, W.M. & McARDLE, M. (1959). Effect of inoculum dilution on spore germination and sporeling growth in a mutant strain of *Aspergillus amstelodami*. *Transactions of the British Mycological Society* 42,235-242.

DAVIS, J.G. (1968). Chemical sterilisation. *Progress in Industrial Microbiology* 8, 141-208.

FENNELL, D.I. (1960). Preservation of fungi. *Biochemical Reviews* 26, 79-141.

FORSYTH, F.R. (1955). The nature of the inhibiting substances emitted by germinating uredospores of *Puccinia graminis* var *tritici*. *Canadian Journal of Botany* 33, 363-373.

FOSTER, J.W. (1939). The heavy metal nutrition of fungi. *Botanical Reviews* 5, 207-239.

FOSTER, J.W., McDANIEL, L.E., WOODRUFF, H.B. & STOKES, J.L. (1945). Microbiological aspects of penicillin. Production of conidia in submerged culture of *Penicillium notatum*. *Journal of Bacteriology* 50, 365-381.

GALBRAITH, J.C.& SMITH, J.E. (1969). Sporulation of *Aspergillus niger* in submerged liquid culture. *Journal of General Microbiology* 59, 31-45.
GILBERT, W.J. & HICKEY, R.J. (1946). Production of conidia in submerged culture of *Penicillium notatum*. *Journal of Bacteriology*. 57, 731-757.
GREENE, H.C. & FRED, E.B. (1934). Maintenance of vigorous mould stock cultures. *Industrial Engineering Chemistry* 26, 1297-1299.
HESSELTINE, A.W. & HAYNES, W.C. (1973). Sources and management of microorganisms for the development of a fermentation industry. *Progress in Industrial Microbiology* 12, 1-46.
INMAN, F.N. (1971). Sterility and sterilisation. Personal Communication.
JOHNSON, M.J. (1972). Techniques for selection and evaluation of cultures for biomass production. *Fermentation Technology Today*. Japan Society of Fermentation pp 473-477.
KAROW, E.O. & WAKSMAN, S.A. (1947). Production of citric acid in submerged culture. *Industrial Engineering Chemistry* 39, 821-825.
LINCOLN, R.E. (1960). Control of stock culture preservation and inoculum build up in bacterial fermentation. *Journal of Biochemical Microbiology, Technology & Engineering* 2, 481-500.
McDADE, J.J., PHILLIPS, G.B., SIVINSKI, H.D. & WHITFIELD, W.J. (1969). Principles and applications of laminar flow design. *Methods in Microbiology* Volume 2, pp 137-168. Edited by J.R. Norris and D.W. Ribbons, London: Academic Press.
MACDONALD, K.D. (1968). Degeneration of penicillin titre in cultures of *Penicillin cyrosogenum*. *Nature* 218, 371-372.
MACDONALD, K.D. (1972). Storage of conidia of *Penicillium chrysogenum* in liquid nitrogen. *Applied Microbiology* 23, 990-997.
METZ, B. & KOSSEN, N.W.F. (1977). The growth of moulds in the form of pellets - a literature review. *Biotechnology & Bioengineering* 19, 281-299.
MEYERS, E. & KNIGHT, S-G. (1958). Studies on the nutrition of *Penicillium roquefortii*. *Applied Microbiology* 6, 174-183.
MEYERS, E. & KNIGHT, S.G. (1961). Studies on the intracellular amino acids of *Penicillium roquefortii*. *Mycologia* 53, 115-122.
MEYRATH, J. & SUCHANEK, G. (1972). Inoculation techniques - effects due to quality and quantity of inoculum. *Methods in Microbiology* Volume 7B, 159-209.
MORTON, A.G. (1961). The induction of sporulation in mould fungi. *Proceedings Royal Society* 153, 548-569.

MOYER, A.J. & COGHILL, R.D. (1946). Penicillin: production of penicillin in surface culture. *Journal of Bacteriology* 51, 57-78.

MUGGLETON, P.W. (1963). The preservation of cultures. *Progress in Industrial Microbiology* 4, 189-214.

O'CALLAGHAN, C. (1977). Personal Communication.

ONIONS, A.H.S. (1971). Preservation of fungi. *Methods in Microbiology* Volume 4,113-151. Edited by C. Booth. London: Academic Press.

RIGHELATO, R.C. (1976). Selection of strains of *Penicillium chrysogenum* with reduced yield in continuous culture. *Journal of Applied Chemical Biotechnology* 26 (3), 153-159.

SAVAGE, G.M. & van der BROOK, M.J. (1946). The fragmentation of the mycelium of *Penicillium notatum* and *Penicillium chrysogenum* by a high speed blender and the evaluation of blended seed. *Journal of Bacteriology* 54, 385-391.

SCOTT, B.R., ALDERSON, T. & PAPWORTH, D.G. (1972). The effect of radiation on the *Aspergillus* conidium. Radiation sensitivity and a 'germination inhibitor'. *Radiation Botany* 12, 45-50.

SEGAL, I.H. & JOHNSON, M.J. (1963). Intermediates in organic sulphate utilisation by *Penicillium chrysogenum*. *Archives of Biochemical Biophysics* 103, 216-226.

SMITH, J.E. (1975). The structure and development of filamentous fungi. *The Filamentous Fungi* Volume 1, pp 1-14. Edited by J.E. Smith and D.R. Berry. London: Academic Press.

SMITH, J.E. & ANDERSON, J.G. (1973). Differentiation in the Aspergilli. *Symposium of the Society for General Microbiology*. 23, 295-337.

STEEL, R., LENZ, C. & MARTIN, S.M. (1954). A standard inoculum for citric acid production in submerged culture. *Canadian Journal of Microbiology* 1, 150-157.

VÉZINA, C., SINGH, K. & SEHGAL, S.N. (1965). Sporulation of filamentous fungi in submerged culture. *Mycologia* 57, 722-736.

VÉZINA, C., SINGH, K. & SEHGAL, S.N. (1968). Transformation of organic compounds by fungal spores. *Advances in Applied Microbiology* 10, 221-268.

VÉZINA, C. & SINGH, K. (1975). Transformation of organic compounds by fungal spores. *The Filamentous Fungi* Volume 1, pp 158-192. Edited by J.E. Smith and D.R. Berry. London: Edward Arnold.

WHITAKER, A. & LONG, P.A. (1973). Fungal pelleting. *Process Biochemistry* (No. 11) 27-31.

PREPARATION, STERILISATION AND DESIGN OF MEDIA

K. Corbett

Beecham Pharmaceuticals, Clarendon Road,
Worthing, Sussex.

INTRODUCTION

A high producing mutant strain of an appropriate organism is the prime requirement for all commercial fungal fermentation processes. Having obtained a good strain it is necessary to obtain the maximum yield of product by the selection of optimum cultural conditions. In his definitive monograph, Pirt (1975) lists five requisite conditions for growth of microorganisms in culture *viz.* an energy source, nutrients to provide the essential materials from which the biomass is synthesised, absence of inhibitors which prevent growth, a viable inoculum, and suitable physico-chemical conditions. Of these parameters, medium design is the sole factor governing the first three and together with the physical condition under which the culture is grown, it contributes significantly to the fifth parameter.

Designing a medium to satisfy the requirements of the organism, in line with the above parameters, may be satisfactory for experimental laboratory purposes, but other factors must be taken into consideration when designing media for biotechnological processes. The objective of such a process is the preparation of product at a cost which enables it to be sold at a profit, therefore, when designing media for industrial processes, economic as well as scientific aspects must be considered. Since the direct cost of the final stage production medium contributes a large proportion to the overall process costs, it is essential to constantly search for cheaper, readily available materials.

The scale of the operation and the type and size of fermenter will vary depending on the end-product. Traditionally, most commercial processes have been operated as a batch or fed-batch process and this is still the case, certainly for the production of secondary metabolites or the over-production of primary metabolites or enzymes. However, the large scale production of single-cell protein has been achieved by continuous processes and some brewing operations are also carried out on a continuous basis. The size of production fermenters has ranged from 10,000 to 200,000 litres. The smaller vessels have been used to produce expensive items required on a limited scale *e.g.* enzymes used in research and for the inter-conversion of products for which there is no easy chemical synthesis. The larger fermenters have been used for both antibiotic and organic acid production particularly when used as raw materials in the food or pharmaceutical industry. However, the development of processes to produce vast quantities of biomass as a source of protein by batch methods has led to the need for larger fermenters and it is reported that a mammoth 4 million litre vessel has been built (Schreier, 1977).

A typical production medium would be a complex heterogeneous mixture with a dry solids content in excess of 10%, therefore, even for conventional fermenters, storage and handling of the medium constituents poses considerable logistic problems. The amount of corn steep liquor, only one of several nutrients, required to batch a 200,000 l fermenter for a penicillin fermentation, would be equivalent to the contents of a road tanker. Operations on this scale place great emphasis on selecting materials which are readily available, reproducible, stable and easy to handle and store. An unwise choice can add considerably to storage, handling or sterilisation cost during medium preparation or to harvesting or extraction cost during product isolation. As a consequence of the scale of operation, the preparation and sterilisation stage can occupy a considerable time during which unwanted chemical reaction may occur with a resultant effect on the final composition of the medium. This possibility must always be considered when scaling up medium from laboratory design to production from plant actuality.

MEDIUM DESIGN

Ideally, before tackling the design of a new medium, it would be advantageous to have a comprehensive knowledge of the primary and secondary biochemical characteristics of the organism. Unfortunately, this is rarely if ever the case. At the initiation of a new biotechnological process, or during the course of its development, one is dealing with a new mutant organism. It is essential to devise suitable growth conditions before conducting more isoteric studies. Medium development therefore must of necessity be an empirical process, the only gauge of success being synthesis of the final product.

The final stage medium is but one of a family of media necessary for a successful development programme. Specialised media are required for mutation, selection, preservation and for seed stages. Moreover, at each stage it is necessary to have available media to test for viability and for contaminating organisms. Successful development of the process entails optimisation of each stage not in isolation but in relation to the overall process. It is pointless to develop one stage of the process to its ultimate efficiency if it does not fit in with the remainder of the process.

The mutation, selection and preservation stages, together with viability and contamination testing, are laboratory scale operations so that in terms of overall cost the medium components are of negligible importance. The bulk of development work goes into improving the seed and final stage media, which are the major cost centres of the process, therefore, further discussion will concentrate on these two parts of the operation.

The seed stages provide a sufficiently large inoculum to ensure rapid growth in the final stage without having to resort to producing unduly large numbers of spores, which would be necessary to directly inoculate the final stage fermenter. In the case of single cell protein production and for the isolation of certain cell-bound constituents, as much biomass as possible is required whereas for the production of secondary metabolites, growth must be limited before the required product is synthesised.

In order to obtain rapid growth in both seed and final stages, media must contain adequate supplies of nitrogen, an energy source, trace elements and specific growth factors which the organism cannot manufacture itself. As much as 15% of the mycelial dry weight may be composed of nitrogen, therefore, he medium must contain up to this amount of a suitable source of nitrogen. Many organisms will use inorganic nitrogen in the form of nitrate or ammonia but even in these cases growth is often stimulated by the addition of suitable organic nitrogenous compounds. The energy source usually consists of a carbohydrate or a lipid component but the carbon skeletons of organic nitrogenous compounds may also contribute. Although the majority of structural materials and components of intermediary metabolic pools may arise from simpler components, there may be a few key metabolites which the organism cannot itself synthesise. The relationship between these growth factors and the growth of many organisms of industrial importance has not been elucidated. Usually these growth factors are added as a part of a complex mixture *e.g.* yeast extract, corn steep liquor; and the nature, much less the amount required, is imperfectly understood. As a consequence of using complex technical grade materials the required trace elements are supplied automatically and it is only the larger requirements for specific elements that has to be met by addition of specific inorganic salts *e.g.* phosphates. As with all medium design, the way to good growth is a balanced medium containing all the above constituents at concentrations appropriate for the specific organism to be cultured.

Designing media for good growth is easier than the subsequent step of obtaining optimum product accretion. The principle of inducing secondary metabolite production is to develop a final stage medium which, after a rapid initial growth phase, will become deficient in one or more of the nutrients. Exhaustion of any nutrient makes balanced growth impossible and brings about biochemical differentiation (Bu'Lock, 1974; Bu'Lock *et al.*, 1974). The nutrient most widely studied for its effect on secondary metabolite production is phosphate, high concentrations of which stimulate vegetative growth and inhibit secondary metabolite production. Martin (1977) has reviewed the regulation of antibiotic formation by bacteria. Secondary metabolites produced by fungi may also be controlled by this nutrient *e.g.* citric acid (Marchesini, 1966), bikaverin (Brewer *et al.*, 1973). This latter fermentation gives a good demonstration of biochemical differentiation in response to phosphate depletion, since not only is ergot alkaloid production induced but there is also a change in cell type from the sphaecelial to the sclerotial form and an increase and change in composition of intracellular lipids (Mantle, Horris & Hall, 1969). Several theories exist to explain how phosphate exerts its effect and these have been reviewed by Martin (1977).

A similar effect to phosphate depletion may be achieved by controlling the levels of carbohydrate, the classic example being penicillin G production. It was realised early in the development of the penicillin fermentation that lactose was a better carbohydrate source than glucose since it was slowly hydrolysed and free glucose in the medium was kept to a minimum. In commercial operations the use of lactose has been replaced by the technique of fed-batch culture whereby it is possible to accurately control the glucose level throughout the fermentation. This procedure limits the rate of substrate utilisation and in this way it is possible to mitigate the effect of catabolite repression (Demain, 1972).

Stimulation of secondary metabolite production by limitation of the nitrogen source is less common. During the production of a nitrogen containing secondary metabolite, limitation of the nitrogen supply may result in cessation of production and an increase in the amount of biomass (Fig. 1) resulting from imbalance of growth and increased storage of materials within the cell. Just as too little nitrogen can be disadvantageous the same may apply to too much nitrogen (Fig. 2). In this instance the C/N balance is the factor that influences the fermentation, allowing controlled growth and antibiotic formation.

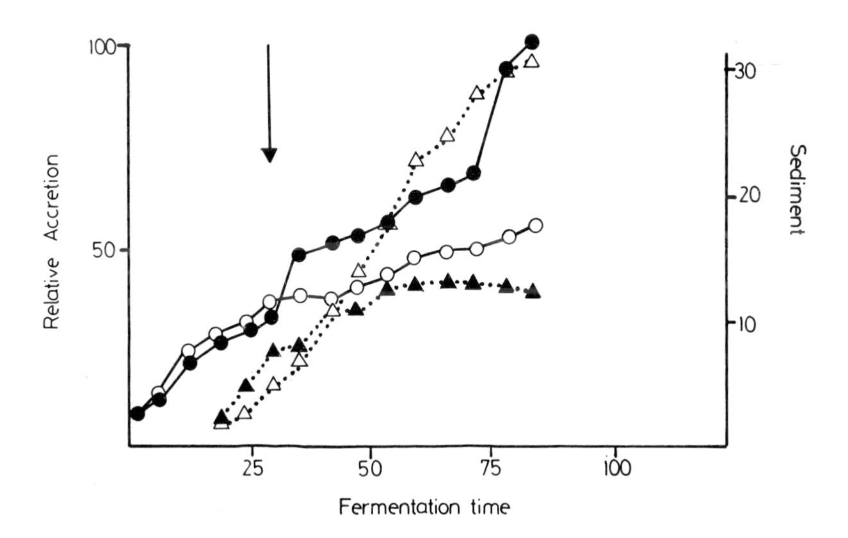

Fig. 1. Effect of nitrogen limitation on growth of *Penicillium chrysogenum* and on production of penicillin G; o———o, biomass (sediment) using standard culture conditions; Δ.......Δ, penicillin G accretion using standard culture conditions, •———• biomass (sediment) using nitrogen limited conditions, ▲......▲ penicillin G accretion using nitrogen limited conditions; the arrow indicates the time at which addition of ammonia was terminated to produce nitrogen limitation.

Dramatic effects can be produced by careful control of the concentration of certain ions. The obvious example is the formation of citric acid by *Aspergillus niger* Production is considerably increased when the medium is deficient in iron and copper. When an impure carbohydrate source is used, such as molasses, it is necessary to remove the iron either

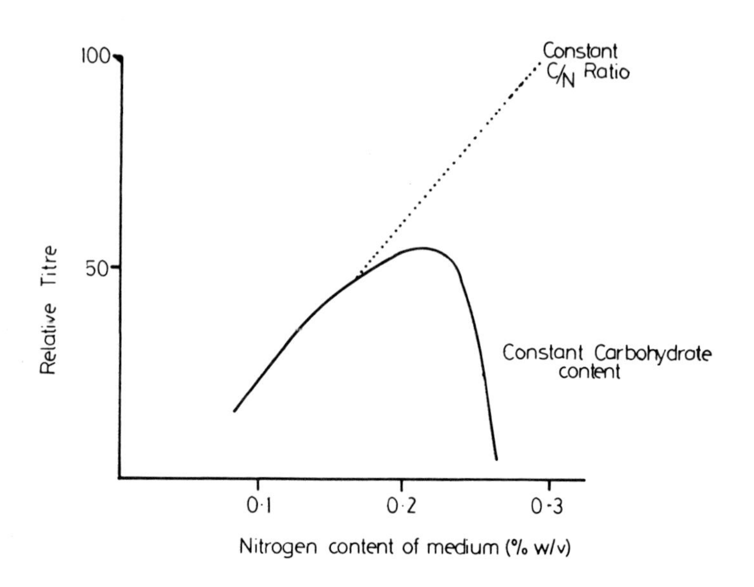

Fig. 2. Effect of carbon/nitrogen ratios on antibiotic production.

by precipitation with ferri-cyanide (Clark, 1962) or by the use of ion exchange resins (Snell & Schweiger, 1949). Similarly *Candida guilliermondii* only over produces riboflavin when the iron content of the medium is restricted (Straube & Fritsche, 1973).

Addition of specific compounds to stimulate secondary metabolite production, as opposed to removing inhibitory substances, has met with limited success. Precursor addition is an obvious possibility although it is usually necessary to have obtained a detailed knowledge of the biosynthetic sequence. Probably the most successful precursor addition *viz.* phenylacetic acid, the side-chain of penicillin G, was discovered well in advance of any knowledge of the biosynthetic sequence. As a component of corn steep liquor and later by separate addition, it transformed the early development of the penicillin fermentation process. Not only did it stimulate penicillin synthesis, but it altered the final product, resulting in the production of penicillin G rather than the aliphatic side-chain penicillins which had previously been produced on Czapek-Dox medium (Moyer & Coghill , 1946). This discovery led to the rapid commercial development of the penicillin G process. A whole range of side-chain precursors have since been shown to be incorporated (Cole, 1966) but phenylacetic acid still remains the most important. Not all precursors are as useful; cysteine and valine, the precursors of the penicillin nucleus are at least as expensive as penicillin and neither will stimulate production on a complex medium. It has been claimed that the D-isomer of valine, the form in which it occurs in the penum nucleus, is even an inhibitor of penicillin accretion (Stevens, Vohra & DeLond, 1954; Demain, 1956), although this finding has not been substantiated (Bycroft *et al.*, 1955).

Apart from precursors of the final product, the literature contains references to a range of stimulating materials being added to media. Surfactants have increased the production of proteases, reputedly by improving permeability and therefore enhancing the availability of important metal ions (Bhumibhamon & Eklund, 1976). Methionine and norleucine have increased the production of cephalosporins while thiosulphate has had a similar effect on penicillin fermentations (Hockenhull, Wilkin & Quilter, 1953), without the reasons being satisfactorily elucidated. Citric acid, when added to ergot alkaloid fermentations at the level of 1%, controls growth and productivity on media containing 30% sucrose (Amici *et al.*, 1967).

The above examples give an indication of the range of compounds which can affect the productivity of particular strains. Medium development is a mixture of carefully planned experiments to cover as many possibilities and combinations as possible and the chance observation which can produce a dramatic effect. The medium must be specifically adapted to each and every mutant, therefore, the job of medium development can never be finished as long as the mutation programme is successful.

Selection of a suitable medium, however, depends on factors additional to optimising the productivity of the organism. Availability, reliability and price of nutrients; transportation, handling and storage costs; ease of preparation and sterilisation, while health and safety considerations must also be taken into account.

The fluctuation of market prices of raw materials is exemplified in Fig. 3, which shows the price changes in three vegetable oils over the period 1974 to 1978. Users of palm kernel oil witnessed an escalation of their process costs in 1974 and 1977 when prices of this commodity were 3 to 4 times the market low price of 1975. Such fluctuations can have severe implications on profitability, therefore, it is advisable to have available alternative nutrients which can be used to alleviate the effects of this type of price variation.

Events remote from the industry can affect the availability and price of nutrients. Firms using ground nut oil in November

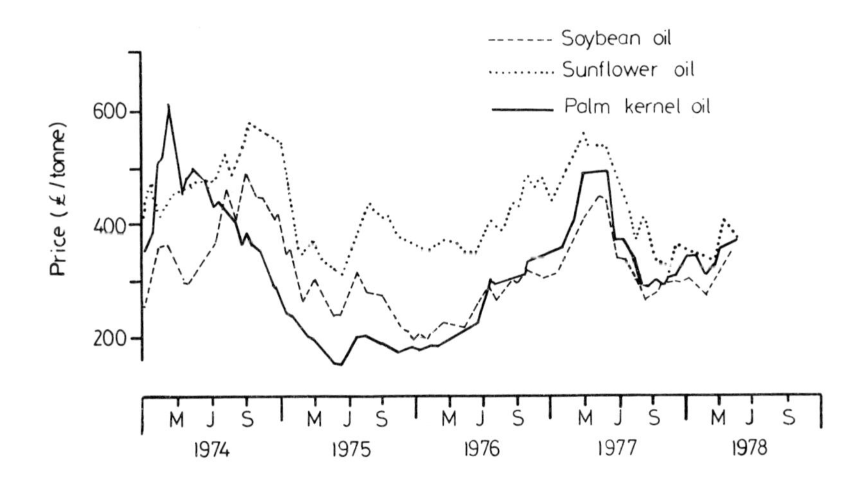

Fig. 3. Fluctuation of the price of three vegetable oils in the period 1974 to 1978.

1973 were subjected to severe price increases over a very short period of time. As a result of the Nigerian civil war, the price of ground nut oil rose from approximately £257/tonne in November 1973 to £500/tonne in January 1974. The industrial development scientist must continually investigate alternative nutrients and keep an eye on the market in order to utilise the cheapest possible source. Obviously the price of the nutrient must be allied with its suitability to the organism; natural oils are complex mixtures and vary considerably in fatty acid composition, therefore, not all are of equal value. Even simple substances like chalk, made by different processes, may result in different productivity as shown in Table 1.

Design of the optimum medium for a biotechnological process therefore depends on three considerations *viz.* satisfying the nutritional demands of the organism; obtaining cheap, economically stable raw material supplies of consistent chemical composition; and selecting the best type of material to ensure smooth operation of the production plant. Since the medium is the major cost centre, it is essential to the economic viability of theprocess that the best medium be developed and that it is kept under constant review to obtain every advantage from changes in the commodity market.

Table 1. Effects of different samples of calcium carbonate on Penicillin G production

Sample	Comparative Titre (%)
Control (Sturcal[1])	100
Omya D5[2]	62.9
Omya Vedar[2]	53.4
Omya Calibrite[2]	71.1
Millicarb[3]	57.9
B.P. Brade	70.0

[1] J & E Sturge Limited

[2] Melbourne Chemical Company

[3] Croxton Garry

PREPARATION OF MEDIA

Included under this heading is the storage, handling and mixing of the media. This may appear to be a rather mundane subject but it is one which is crucial to the successful operation of any biotechnological process. It is an area which may not require erudite scientific thought, but nevertheless it is fraught with technical problems which must be solved if the operation is to be performed efficiently.

Storage of bulk materials requires special attention. Powders must be kept in dry conditions otherwise materials such as chalk and phosphate can become rock-like in composition. Protein solids when damp can rapidly form glutinous lumps which make subsequent handling and sterilisation unpredictable. Many bulk liquids contain a high solids content and must be kept warm to prevent them solidifying *e.g.* glucose and corn steep liquor. If a glucose solution solidifies in tanks and pipes it is expensive and time-consuming to return it to the liquid state. On the other hand if the storage temperature is set too high the material may be degraded with subsequent impairment of the ensuing fermentation. Handling solid materials poses problems in relation to health and safety. More and more employers are insisting that when powders are handled, the operators must wear protective clothing and breathing apparatus. Even with well designed over-suits and fresh air hoods, freedom of movement is restricted and the operator becomes hot and uncomfortable particularly on a fermentation plant, therefore, prolonged use of such equipment is unpopular. When designing media one must bear such considerations in mind.

Whereas it may be reasonable to expect a highly trained technician to accurately follow a medium recipe in a laboratory, the use of unskilled labour, handling unfamiliar materials in a hot, humid plant, possibly in the middle of the night, increases the chances of mistakes arising, therefore careful supervision is essential. Operators must be trained to follow batching instructions implicitly. Incorrect order of batching may lead to 'lumpy' medium which may result in some of the nutrients being less freely available as well as hindering sterilisation. Incorrect batching may also lead to unwanted chemical reactions with resultant loss of nutrients.

Preparing the medium is the least glamorous part of the operation but if the fruits of good medium design are to be bathered and the process is to run smoothly, then medium preparation in the production plant is the corner-stone of the whole operation.

Sterilisation

Theoretically it should be possible to sterilise media by mechanical removal of unwanted organisms or by *in situ* killing by heat, chemical agents or electromagnetic radiation. The large volumes involved have precluded the use of radiation methods and the hazardous nature of potentially suitable chemical agents, together with the necessity of removing them before inoculation, has severely limited their use. Totally soluble media, soluble feeds for fed-batch processes or water for making up to volume may all be sterilised by filtration. However, in our experience, present day membrane filters are too fragile, therefore, until more reliable and less costly systems are available, they will not supercede traditional means of medium sterilisation. The universal method for sterilising typical heterogeneous industrial media is steam heating under pressure.

There are four reasons for removing unwanted organisms and starting with sterile media:

a) they may change the environmental conditions *e.g.* they may change the pH away from the desired value or they may reduce availability of oxygen;
b) they may use up expensive nutrients - an extreme case occurs in penicillin fermentations which have been contaminated with *Paecilomyces* which can utilise phenylacetic acid (an

expensive item) faster than *Penicillium*;
c) they may destroy the required product *e.g.* contamination of a penicillin fermentation with a β-lactamase producer can result in all the penicillin being destroyed in a matter of hours;
d) the sheer mass of organisms may cause viscosity problems and interfere with subsequent harvesting of the medium. This is particularly the case when yeast or filamentous organisms are the contaminants and a filtration system is being used to harvest the product.

Bacterial spores are the most heat resistant type of organism known and moisture is a necessary adjunct for efficient sterilisation. Oil droplets and conglomerations of organic materials, so common in industrial media, can seriously affect the efficiency of steam sterilisation. Destruction of encapsulated spores depends on the diameter of the particles, the thermo-conductivity of the material, permeability to steam and water and the relative humidity in the immediate vicinity of the spores. Relative humidities as low as 20% may occur within oil droplets or organic conglomerates, resulting in kill rates 2 to 3 order of magnitude lower than in the aqueous phase of the medium (Fedossejev & Kampe-Nenm, 1976). Consequently, it may be necessary to prolong sterilisation after the death of all the spores in the liquid phase and to determine empirically the length of time necessary to obtain sterile medium.

The necessity to heat a complex chemical mixture such as a culture medium at high temperatures for periods exceeding 20 min is not without its disadvantages. A number of important components may be heat labile and because of the complex nature of the mixture, a whole range of chemical reactions may occur, one of the best known being the Maillard or browning reaction between reducing sugars and amines which can result in the destruction of a considerable portion of the added sugars. The dramatic effect on glucose concentrations during sterilisation is shown in Table 2. The longer the time at high temperature, the greater the destruction of glucose and the lower the production of antibiotic.

The extent of the changes in the medium will depend not only on the temperature but also on other parameters, including the duration of heating, pH, composition of the medium and presence of oxygen. An example of changes in the medium under different conditions is exemplified in Table 3. The amount of added nitrogenous material, mainly protein, which remains in solution after the sterilisation procedure, depends on the pH of the medium, the time of heating and the addition of trace elements.

Table 2. Effect of sterilisation time on glucose concentration and antibiotic accretion rate

Time at 121°C (min)	% of added glucose remaining	Relative accretion rate
60	35	90
40	46	92
30	64	100

Table 3. Effect of sterilisation conditions on soluble nitrogen content of medium

	Sterilisation conditions		% of total nitrogen in soluble form
pH	Time at 121°C (min)	Addition of trace elements	
6.4	20	Yes	33
6.4	40	Yes	42
7.0	20	Yes	67
7.0	40	Yes	68
6.4	20	No	45

Methods of Sterilisation

The effect of temperature on the reaction rate constant, k, for the heat inactivation of spores can be expressed by the Arrhenius equation,

$$\ln k = C - \frac{E}{RT}$$

E = activation energy
C = constant
R = Bolzman's constant

Similarly, the rate constant for the destruction of medium components will depend on temperature. However, the activation energy for many chemical reactions including destruction of growth factors is in the range of 10-30 kcals/mole *e.g.* hydrolysis of casein 20.6 kcals/mole; destruction of riboflavin 20.5 kcals/mole, whereas the activation energies for the heat destruction of spores lies in the range of 50-100 kcals/mole. If one plots ln k against 1/T, the slope of the line is equal to -E/R and since R is a constant, the slope depends on the activation energies (Fig. 4). Reactions with a high energy of activation have a steeper slope, therefore the rate constant rises faster with increasing temperature than reactions with a low energy of activation. Consequently, raising the temperature increases the rate of spore destruction more quickly than the destruction of nutrients. The practical application of this phenomenon is the flash or high-temperature-short-time steriliser comprising a series of continuous flow heat-exchangers and a holding coil. Medium is rapidly heated, held at sterilisation temperature for a short period of time and then rapidly cooled. Traditionally, plate type heat exchangers have been used but they suffer from the disadvantage that the plates are separated by gaskets which constantly require replacing and should one fail in operation the non-sterile and sterile streams can mix. Modern heat exchangers are of the double spiral type in which the two streams of liquid or liquid and steam are separated by a continuous stainless-steel division. In operation the system is initially sterilised by cycling water through the system to establish operating temperatures and pre-sterilise the equipment. Once this has been achieved cold medium is introduced into the first heat exchanger, or the economiser, which allows the incoming liquid to receive heat from the existing sterile stream. In this way, the incoming liquid is pre-heated and the outgoing sterile stream is cooled with considerable savings in energy; in a well designed system, 60% of heat can be reused. The medium passes to the main heat exchanger in which steam is passed through the second spiral thus bringing the medium up to the sterilisation temperature of 140-160°C. Once at temperature, the medium is held in a lagged serpentine coil for the desired time, usually 2 to 4 min. The hot medium is returned to the economiser and finally to a water cooled exchanger before passing to a pre-sterilised fermenter.

A temperature profile of medium sterilisation in a flash steriliser is shown in Fig. 5. The medium is above 100°C for less than 4 min. Compare this with the temperature profile for batch sterilisation (Fig. 6). The length of time the medium is at high temperature depends on the volume of the broth and the actual method employed; with larger volumes the heat-up time and cool-down times are extended. In order to speed up the turn round time of fermenters, a separate batch steriliser is often used, allowing the fermenter to be prepared and pre-sterilised concurrently with sterilising the medium. To minimise heat losses the steriliser is lagged, therefore, the medium will cool slowly. Moreoever, steam pressure is often applied to the vessel to force it through

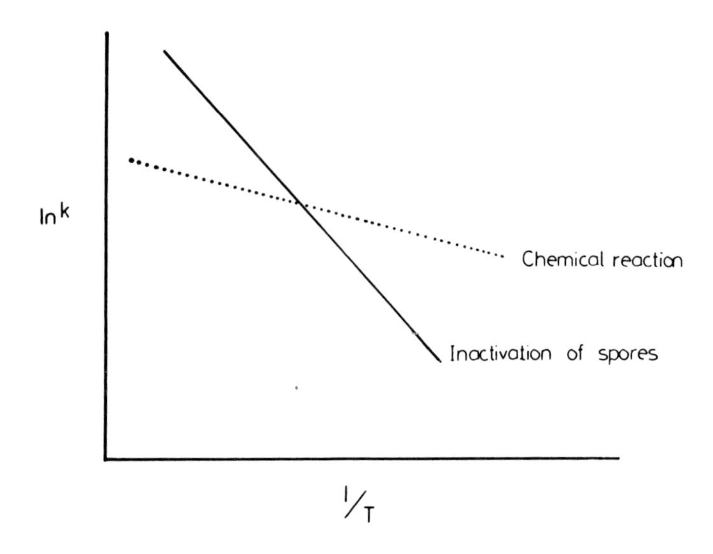

Fig. 4. Effect of temperature on the rate constants for chemical reactions and inactivation of spores.

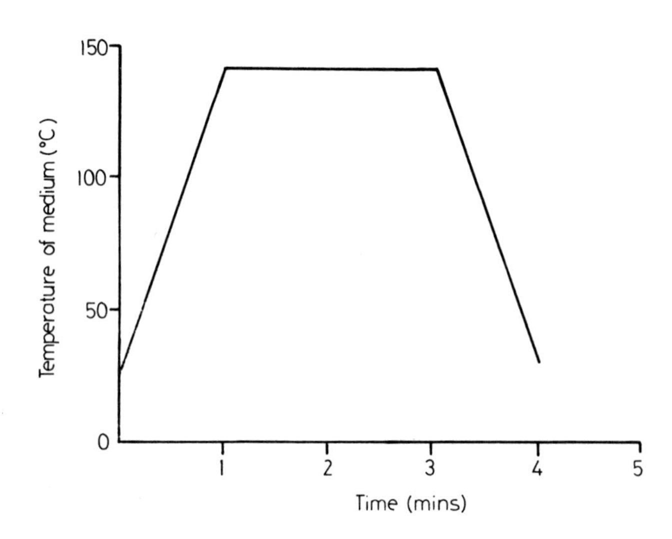

Fig. 5. A typical temperature profile of culture medium subjected to continuous sterilisation.

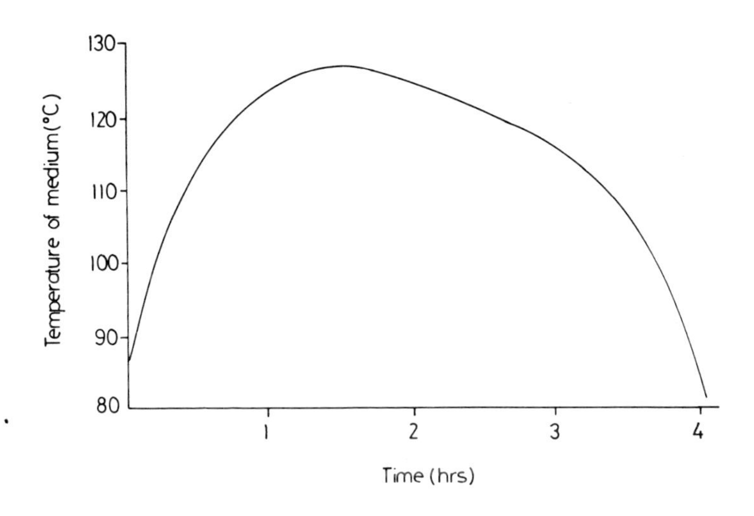

Fig. 6. A typical temperature profile of culture medium subjected to batch sterilisation.

the transfer system to the fermenter in which case, part of the medium remains near to sterilisation temperature until it has all been transferred. It is not uncommon with large vessels that the medium is above 100°C for more than 4 h.

Despite the obvious advantages in terms of nutrient integrity, flash sterilisation has not been widely used on the large scale because of the high initial cost and the complexity of the equipment. However, the improvement offered by the continuous welded double-spiral heat exchanger over the plate system is increasing the usage of flash sterilisation in biotechnology.

CONCLUSIONS

The key to obtaining the full potential of a good mutant lies in optimising the design of media for each phase of the process. Once a good medium has been designed it is essential to ensure that the preparation of the medium is carried out in a manner which is accurate and does the least damage to the medium. Similarly, during the sterilisation the medium must be given the mildest treatment conducive with sterility to ensure the integrity of the original design. At all stages cost plays a limiting role in deciding the composition and subsequent treatment of the medium since the successful development of any biotechnological process must ultimately depend on its economic viability in relation to other methods of producing the same product.

REFERENCES

AMICI, A.M., MINGHETTI, A., SCOTTI, T., SPALLA, C. & TOGNOLI, L. (1967). Ergotamine production in submerged culture and physiology of *Claviceps purpurea*. *Applied Microbiology* 15, 597-602.

BHUMIBHAMON, O. & EKLUND, E.(1976). The effects of some surfactants on the production of acid protease by *Aspergillus awamori* and *Aspergillus phoenicis*. *Abstracts of the 5th International Fermentation Symposium*. (Ed. H. Dellweg), pp 154. Berlin: Verlag Versuch-und Lehranstalt fur Spiritusfabrikation und Fermentationstechnologie.

BREWER, D., ARSENAULT, G.P., WRIGHT, J.L.C. & VINING, L.C. (1973). Production of bikaverin by *Fusarium oxysporum* and its identity with lycopersin. *Journal of Antibiotics* 26, 778-783.

BU'LOCK, J.D. (1974). Secondary metabolism of microorganisms. *Industrial Aspects of Biochemistry* (9th Meeting of the Federation of European Biochemical Societies, Dublin, April, 1973). (Ed. B. Spencer), pp 335-346. Amsterdam: North Holland/American Elsevier.

BU'LOCK, J.D., DETROY, R.W., HOSTALEK, Z. & MUNIN-AL-SHAKARCHI, A. (1974). Regulation of secondary biosynthesis in *Gibberella fujikuori*. *Transactions of the British Mycological Society* 62, 377-389.

BYCROFT, B.W., WELS, C.M., CORBETT, K., MALONEY, A.P. & LOWE, D.A. (1975). Biosynthesis of penicillin G from D- and L- (14_C)- and (α-3_H)-valine. *Journal of Chemical Society Chemical Communications* pp. 923-924.

CLARK, D.S. (1962). Submerged citric acid fermentation of ferricyanide-treated cane molasses. *Biotechnology and Bioengineering* 4, 17-21.

COLE, M. (1966). Microbial synthesis of penicillins. *Process Biochemistry* 1,334-338.

DEMAIN, A.L. (1956). Inhibition of penicillin formation by amino-acid analogs. *Archives of Biochemistry and Biophysics* 64, 74-79.

DEMAIN, A.L. (1972). Cellular and environmental factors affecting synthesis and excretion of metabolites. *Journal of Applied Chemistry and Biotechnology* 22, 345-362.

FEDOSSEJEV, K.G. & KAMPE-NEMM, A.A. (1976). Sterilisation of nutrient media suspensions. *Abstracts of the 5th International Fermentation Symposium*. (Ed. H. Dellweg). pp 43. Berlin: Verlag Versuch-und Lehranstalt fur Spiritusfabrikation und Fermentationstechnologie.

HOCKENHULL, D.J.D., WILKIN, G.D. & QUILTER, A.R.J. (1955). Penicillin. *British Patent* 730,185.

MANTLE, P.G., MORRIS, L.J. & HALL, S.W. (1969). Fatty acid composition of sphaecelial and sclerotial growth forms of *Claviceps purpurea* in relation to the production of ergoline alkaloids in culture. *Transactions of the British Mycological Society* 53, 441-447.

MARCHESINI,A. (1966). Effects of chelating agents on the solubility of soil phosphorus. IV. Relation between phosphorus availability and the growth of *Aspergillus niger*. *Annali di Microbiologia ed Enzimologia* 16, 147-151.

MARTIN, J.F. (1977). Control of antibiotic synthesis by phosphate. *Advances in Biochemical Engineering* 6, 105-127.

MOYER, A.J. & COGHILL, R.D. (1946). Penicillin. X. The effect of phenylacetic acid on penicillin production. *Journal of Bacteriology* 53, 329-341.

ROBERS, J.E., ROBERTSON, L.W., HORNEMANN, K.M., JINDRA, A. & FLOSS, H.G. (1972). Physiological studies on ergot. *Journal of Bacteriology* 112, 791-796.

SCHREIER, K. (1977). The 4,000 m^3 fermenter for petro-protein production. *Abstracts of the 4th FEMS Symposium, Vienna* March, 1977. pp 831.

SNELL, R.L. & SCHWEIGER, L.B. (1949). Production of citric acid by fermentation. *United States Patent* 2,492,667.

STEVENS, C.M., VOHRA, P. & DELONG, C.W. (1954). Utilisation of valine in the biosynthesis of penicillins. *Journal of Biological Chemistry* 211, 297-300.

STRAUBE, G. & FRITSCHE, W. (1973). Influence of iron concentration and temperature on growth and riboflavin overproduction of *Candida guilliermondii*. *Biotechnology and Bioengineering, Symposium No. 4* pp 225-231.

GENETIC MODIFICATION OF FILAMENTOUS FUNGI

C. Ball

Glaxo Operations U.K. Limited,
Ulverston, Cumbria

INTRODUCTION

Once a microorganism has been chosen for use in industrial fermentations the ease with which process economics can be improved by genome modification will depend on the type of organism and the type of fermentation end-product. The former is usually a bacterium, actinomycete or fungus. The type of end-product can range from primary and secondary metabolites such as amino acids and antibiotics through primary and secondary macromolecules such as enzymes and virus-like particles, to whole organisms as harvested in the microbial protein industry.

The economy of the process can be improved in several ways and these will be discussed in this paper. Productivity can be improved. Qualities other than productivity such as substrate utilisation efficiency can be improved and novel compounds related to the original product can be produced. The techniques used include mutation and recombination to increase the number of genetic variants followed by screening and selection to detect the variants preferred. Great care is required in handling strains. Undesirable spontaneous mutations can sometimes occur at a high rate and/or be selected for, giving rise to degeneration of a strain's industrial performance.

STRAIN IMPROVEMENT

A colony that appears on agar medium, following plating out of spores, cells or small hyphal fragments, can be defined as a strain. A colony consists of a population of nuclei most of which are identical. Some may differ however, due to spontaneous mutation during growth of the colony or to nuclear heterogeneity in the original propagule. Clearly, a strain is a heterogeneous population of nuclei. To maintain it the qualitative structure of the nuclear population must be maintained at least with respect to those nuclei that influence important characteristics. In practice, considerable thought must be given to the method of propagating strains. For example, if a colony is subcultured by propagating a small region the new nuclear population could be qualitatively different from that constituting the original strain (Sermonti, 1969). Therefore, not surprisingly the industry has always placed great emphasis on strain preservation techniques with the main object of retaining viability and productivity potential of the preserved biological material. A good example is the work of MacDonald (1972) showing liquid nitrogen to be superior to storage of slopes or slope-to-slope transfer for preserving the productivity potential of *Penicillium chrysogenum*.

However, despite elaborate preservation and propagation methods, a strain has to be grown in a large production fermenter in which the chances of genetic change are very high through spontaneous mutation and selection. The chance of a high rate of spontaneous mutation is probably greater when the industrial strains in use have resulted from many years of mutagen treatment. Answers to such instability problems are often found without knowing the exact cause and without knowing the exact reason why the answer succeeds. An example of this empirical approach in identifying relatively stable strains is the taking of samples throughout shake flask, pilot plant and main plant fermentations, followed by testing them for titre in shake-flasks to check for consistency of results.

Some possible causes of genetic instability are:

1. Diploidy or partial diploidy i.e. aneuploidy (Roper, (1973).
2. Partial chromosome duplication (Roper, 1973).
3. Unstable point mutation e.g. base pair (Drake, 1973).
4. Autonomous cytoplasmic factors e.g. mitochondrial DNA (Preer, 1971).
5. Insertional sequences (Starlinger & Saedler, 1976).

Some possible cures are:

1. Improved preservation techniques (MacDonald, 1972).
2. Alteration of growth medium (Righelato, 1976).
3. Reduced number of nuclear divisions (Sermonti, 1969).
4. Re-isolation, using morphological correlations if possible (Backus & Stauffer, 1955).

5. Selection of anti-mutator genes (Drake, 1973).
6. Selection of stable diploids (Elander, 1967).

Note that these cures may operate by reducing either the incidence of instability or the selective advantages of certain types.

An unstable genetic system that has been mentioned in a previous paper (Ball, 1973) will serve here to illustrate cause and cure of instability. This was a morphologically unstable strain of *P. chrysogenum* that also exhibited instability for penicillin productivity. Its reduced productivity correlated with certain types of morphological variant in the population on agar medium. A dense-sporing, morphologically stable strain was selected that had improved preservation characteristics and a reduced penicillin titre which it retained after slope-to-slope transfer. Further mutation, recombination and selection of this strain improved the penicillin productivity beyond that of the original strain. One of the conclusions from this work was that selection for morphological characteristics correlated with selection for titre stability. However, further questions were asked about the cause of the initial instability and a nuclear determination was established following hybridisation of the unstable to the stable strain. Aneuploidy was ruled out because the heterozygous segregants expected on this hypothesis were not detected. Cytoplasmic causes were excluded by the heterokaryon test (Preer, 1971). Partial chromosome duplication was favoured as an explanation because of the superficial analogy with similar instability in *Aspergillus nidulans* (Roper, 1973). Subsequent work by Righelato (1976) established that the stable strain and its derivatives were unstable in glucose-limited chemostat culture at high growth rates, an effect not observed with other limiting substrates such as phosphate. This is a good example of how alteration of growth medium can cure instability without detrimental effect on the relative specific production rate.

GENOME MODIFICATION TO IMPROVE PRODUCTIVITY

As already said the strategy employed to increase productivity depends on the organism and product. The simpler problems are those presented by the task of improving yields of some primary metabolites and some macromolecules, particularly if produced by bacteria (Clarke, 1976). More complex problems occur when trying to improve the yield of such fermentation products as the secondary metabolites produced by fungi and actinomycetes. This paper will mainly consider these. Nevertheless, some general points are worth making about industrial production of primary metabolites and enzymes. First, many of the enzymes of industrial significance are fungal catabolic ones such as proteases and carbohydrate-degrading enzymes. Despite the extensive literature (Demain, 1971) on methods of deregulating catabolic enzymes by removing induction or feedback control or catabolic control, very few of these techniques have been applied consciously in the industrial field (Johnston, 1975). In contrast, the planned deregulation of the anabolic synthesis of primary metabolites such as amino acids and nucleotides has been outstandingly successful in industry. A comprehensive review of the use of auxotrophs and analogues to derepress and remove inhibition in amino acid production has been presented by Nakayama *et al.* (1976).

For secondary metabolites such as antibiotics, Demain (1971) has emphasised the possible regulatory influences analogous to those that control primary metabolism. Examples of supposed induction, feedback control and catabolite control are quoted for secondary metabolic pathways as well as for the pathways leading to formation of primary metabolite building blocks for the appropriate secondary metabolites. Primary metabolites may also act as inducers and derepressors of secondary metabolism (Drew & Demain, 1977). A major difficulty in increasing the production of a secondary metabolite is determining what may limit production in a given strain (see Fig. 1). This lack of knowledge has led to a variety of empirical tricks in the endeavour to increase productivity (Demain, 1973). Such tricks can be categorised by the general headings of screening and selection. These two techniques can be distinguished by their contrasting difference: in a selection system all rare strains grow while the rest do not, whereas in a screening system all strains grow but certain colonies are chosen because of desirable characteristics.

Examples of screening systems are:

1. Titre-testing in shake-flasks under a variety of conditions that may include titre-suppressing compounds.
2. Titre-testing on agar, i.e. looking for size of zone of bacterial inhibition around colonies, or selecting non-producing mutants and reverting them to production.
3. Identifying morphological mutants on agar medium.
4. Screening for auxotrophic mutants by replica plating on agar medium.

Examples of selection techniques are:

1. Resistance to toxic analogues of metabolites.
2. Detoxification of toxic precursor by formation of more secondary metabolite.

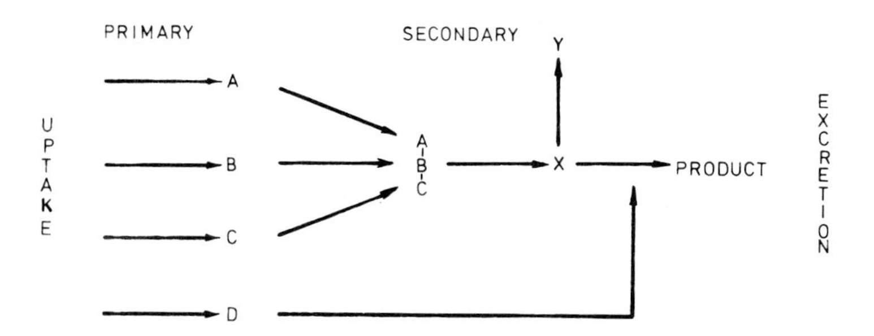

Fig. 1. Various types of metabolic control may be exerted on this pathway (e.g. induction, feedback and catabolic control). Also, energy is required at various steps. The type of control and demand for energy can vary during the fermentation thus complicating the elucidation of steps limiting product formation. If the product is penicillin G, then A = α amino-adipic acid, B = cysteine, C = valine, A-B-C = tripeptide, D = phenylacetic acid, X = isopenicillin N and Y = 6 amino-penicillinac acid. Analysis of control changes that have taken place during empirical strain selection to improve penicillin productivity have recently been discussed by Martin (1978).

---

3. Detoxification of a metabolic inhibitor by direct reaction with the secondary metabolite.
4. Auxotrophic enrichment techniques and autotroph reversion.
5. Resistance of an organism to its toxic secondary metabolite.

Using these techniques mutants with enhanced productivity can sometimes be detected. Demain (1973) gives several accounts of such methods but does not deal with methods of generating genetic variants for such screening and selection. The techniques of mutation and recombination are used to achieve this but in the past by methods that were extremely empirical. However, more elegant methods of carrying out mutation and recombination are clearly being developed. This is particularly exemplified by directed mutagenesis (Godfrey, 1974) and by genetic engineering in prokaryotes such as bacteria and actinomycetes (Murray, 1976; Hopwood, 1978). Nevertheless, the general use of these techniques in improving the yield of secondary metabolites is impeded by several restraints, notably the need for detailed knowledge of the biosynthetic pathway and the enzyme steps involved (Fig. 1). Synthesis of a secondary metabolite is not determined by a single gene and, often, all the genes do not exist in a contiguous operon-like structure even in prokaryotic actinomycetes (Chater, 1978). Thus, we must consider more realistic, immediate approaches to the problem of yield improvement of secondary metabolites.

A major problem in mutagenesis is the evaluation of optimum mutagen and dose. Alikhanian (1962) has indicated that high survival levels are preferable for inducing titre-positive mutations in highly developed industrial strains but not in less developed strains. These are plausible conclusions that now need supporting by similar work by others preferably stating the details of experimental design. Other problems in industrial mutagenesis have been discussed by Sermonti (1969) and Johnston (1975).

A major issue in recombination studies is the relevance of genetic mapping in productivity improvement. Calam, Daglish & McCann ((1976) highlighting this stressed the time factor in strain improvement. There are several answers to that criticism. One is that improved strains can be hybridised with well mapped strains, the time factor permits this while the rate of progress in strain improvement by standard techniques declines. Another answer is that knowledge of the genetic map enables an understanding of the basic genetics of the organism and determination of the principles and strategies to be employed in empirical recombination (breeding) programmes.

Of interest in regard to breeding for productivity improvement is that the experience of workers with fungi (Ball, 1973) and actinomycetes (Hošťálek, Blumauerová & Vaněk, 1974) has led to preference for certain conditions. These include:

1. If possible, parent strains to be different, i.e. separated by several steps of mutation and selection for increased yield. One can only breed from differences (Fig. 2).
2. Parent strains to have similar yields if possible.
3. Markers not to be titre-modifying.
4. Recombinants to be readily selected or induced (Fig. 3 and Upshall, 1978).

The rationale used in breeding strains is illustrated in Fig. 4. In unsuccessful attempts to improve the yield of fermentation product by breeding, many of these conditions have not been observed. In addition, biometrical techniques should be more widely applied in both mutation and recombination studies (Caten & Jinks, 1976). Such techniques allow estimation of the variable genetic and environmental components and can help prevent erroneous deductions when interpreting biological variability.

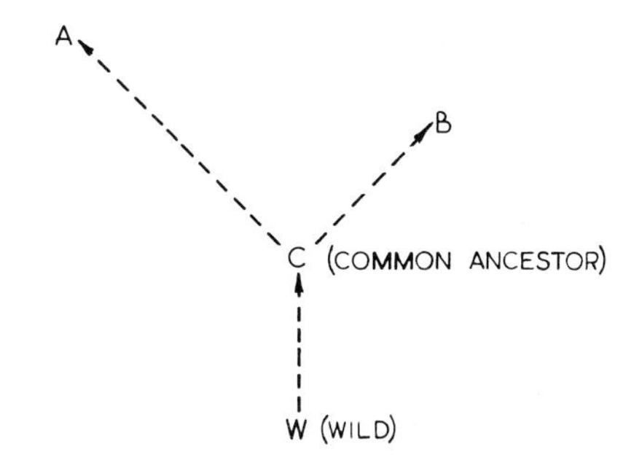

SISTER CROSS eg. A x A
ANCESTRAL CROSS eg. A x W
DIVERGENT CROSS eg. A x B

Fig. 2. The dotted lines indicate serial mutation and selection of haploid strains for enhanced product formation. In carrying out all crosses it is usually necessary to introduce different markers, such as auxotrophs, into the two strains to be crossed. In industrial breeding ancestral and divergent line crosses are often preferred (see text) but rationales also exit for sister strain crosses notably the isolation of "homozygous" diploids in parasexual crosses (Elander, 1967).

GENOME MODIFICATION TO IMPROVE QUALITIES OTHER THAN PRODUCTIVITY

There are many strain qualities other than productivity improvement that can help reduce process costs. These characteristics can affect any stage of the process, i.e. pre-fermentation, fermentation, post-fermentation or the type of end-product. For the pre-fermentation stages, qualities such as strain stability on preservation and propagation have already been mentioned. Simple qualities such as enhanced sporulation (Calam *et al.*,1976) or growth rate on agar can also help to make the overall process more efficient (Hamlyn & Ball, 1978). During fermentation, qualities such as substrate utilisation efficiency and morphology in submerged culture can help reduce costs, the latter by influencing factors such as foaming and power input (Elander, 1967). In the post-fermentation stages the quality of the organism and of the fermentation broth is important. For example, strains with well developed hyphal morphology might be more readily filtered (Ball *et al.*, 1978). Extraction of the final product can be aided by the absence of unwanted compounds in the broth (Backus & Stauffer, 1955).

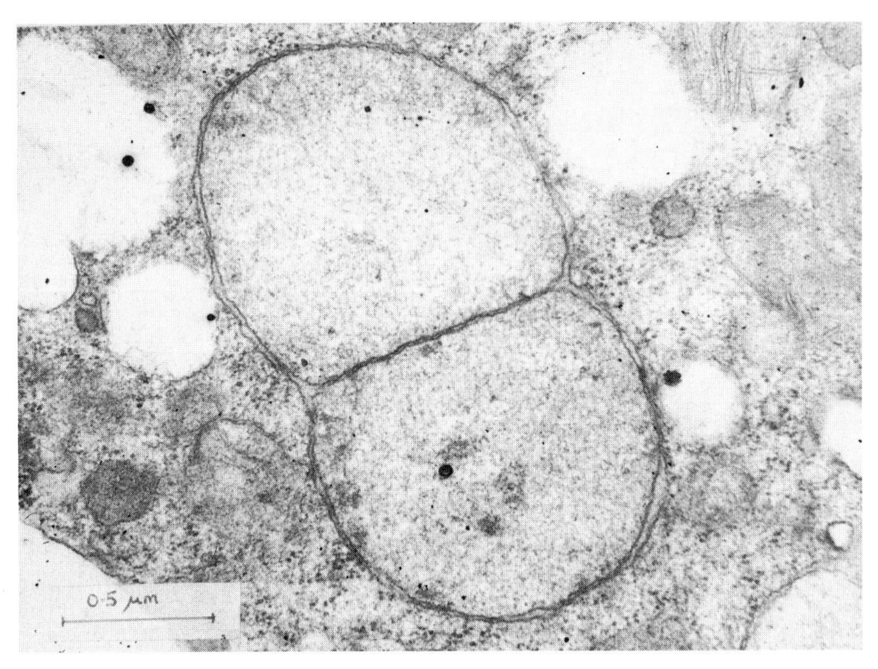

Fig. 3. The adhesion of nuclei following fusion of protoplasts of *Cephalosporium acremonium* is shown. Hamlyn & Ball (1978) showed that this technique was a reliable method of crossing strains of this organism. The genetic evidence indicated that following nuclear fusion recombination and segregation to the haploid condition could take place readily.

Also, in regard to the nature of the final product the history of antibiotic production is full of examples of improved variations on the basic type of molecule (Sermonti, 1969). Many such variations are accomplished by genetic change. It has been suggested by Hopwood (1978) that the new technique of genetic engineering could help to enhance the range of new fermentation products. With fungi protoplasts are likely to figure extensively in such work (Peberdy, 1978; Hinnen, Hicks & Fink, 1978).

As in the instance of productivity improvement, the techniques used to bring about the above changes have been mutation, recombination, screening and selection. On the whole, mutation and screening have outweighed the others but recombination is likely to become increasingly important because when compared with mutation experiments fewer colonies may

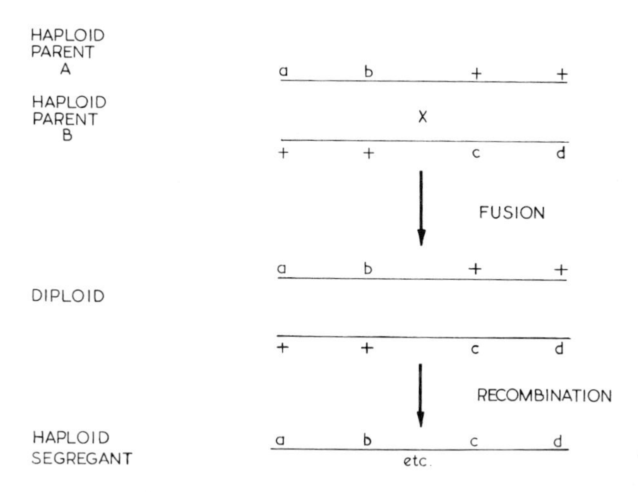

Fig. 4. The breeding rationale is essentially similar when using the sexual and parasexual cycles, one difference being that the diploid state is transient in the former and relatively stable in the latter with a few exceptions (see Fig. 3). The mutants a, b, c and d could be different productivity increasing mutations that interact in various ways. Many of these interactions in haploid recombinants may not give productivity improvements but some may do so. Alternatively, certain of the mutant combinations could determine other desirable fermentation properties (see text) and these might be readily combined with the productivity increasing mutations.

have to be screened. This is because the quality in question can often be identified in the strain to be used in hybridisation (see Fig. 4). The above considerations do not exclude the possibility of producing new qualities by empirical recombination. For example, two foaming actinomycete strains, when hybridised, produced non-foaming recombinants (Sermonti, 1969). Nevertheless, if such techniques seem to a biochemist to be too empirical, the method described as mutational biosynthesis (Nagaoka & Demain, 1975), using defined feeding of alternative intermediates to idiotrophs (mutants of secondary metabolism), will probably seem more interesting.

CONCLUSION

Some of the problems and achievements in genome modification and stability of industrial microorganisms have been discussed mainly by reference to a number of reviews. We have noted a slow movement away from the extreme empiricism that characterised the early days of the fermentation industry. Fundamental studies of the genetics of fungi and other microorganisms provide a background of ideas for empirical solution of industrial problems and have indeed contributed directly to progress in industrial strain selection.

REFERENCES

ALIKHANIAN, S.I. (1962). Induced mutagenesis in the selection of microorganisms. *Advances in Applied Microbiology* 4, 1-50.

BACKUS, M.P. & STAUFFER, J.F. (1955). The production and selection of a family of strains of *Penicillium chrysogenum*. *Mycologia* 47, 429-463.

BALL, C. (1973). The genetics of *Penicillium chrysogenum*. *Progress in Industrial Microbiology* 12, 47-72.

BALL, C., LAWRENCE, A.J., BUTLER, J.W. & MORRISON, K.B. (1978). Improvement of amyloglucosidase production following genetic recombination of *Aspergillus niger* strains. *European Journal of Applied Microbiology and Biotechnology* 5, 95-102.

CALAM, C.T., DAGLISH, L.B. & McCANN, E.P. (1976). Penicillin: tactics in strain improvement *2nd International Symposium on the Genetics of Industrial Microorganisms* (Ed. K.D. Macdonald), pp. 273-287. London U.K. and New York U.S.A.: Academic Press.

CATEN, C.T. & JINKS, J.L. (1976). Quantitative genetics. *2nd International Symposium on the Genetics of Industrial Microorganisms* (Ed. K.D. Macdonald), pp. 93-111. London U.K. and New York U.S.A.: Academic Press.

CHATER, K.F. (1978). Some recent developments in *Streptomyces* genetics. *3rd International Symposium on the Genetics of Industrial Microorganisms* In Press. American Society for Microbiology.

CLARKE, P. (1976). Mutant isolation. *2nd International Symposium on the Genetics of Industrial Microorganisms* (Ed. K.D. Macdonald), pp. 15-28. London U.K. and New York U.S.A.: Academic Press.

DEMAIN, A.L. (1971). Overproduction of microbial metabolites and enzymes due to alternation of regulation. *Advances in Biochemical Engineering* (Eds. T.K. Ghose, A. Fiechter and N. Blakeborough) 1, 113-142. Berlin and Heidelberg, Germany and New York U.S.A.: Springer-Verlag.

DEMAIN, A.L. (1973). Mutation and the production of secondary metabolites. *Advances in Applied Microbiology* 16, 177-202.

DRAKE, J. (1973). The genetic control of mutation. *Genetics* (Supplement) 73, 1-205.

DREW, S.W. & DEMAIN, A.L. (1977). Effect of primary metabolism on secondary metabolism. *Annual Review of Microbiology* 31 343-356.

ELANDER, R.P. (1967). Enhanced penicillin biosynthesis in mutant and recombinant strains of *Penicillium chrysogenum. Abhandlungen Akademic Der Wissenschaften* 4, 403-423.

GODFREY, O.W. (1974). Directed mutagenesis in *Streptomyces lipmanii. Canadian Journal of Microbiology* 20, 1479-1485.

HAMLYN, P. & BALL, C. (1978). Recombination studies in *Cephalosporium acremonium. 3rd International Symposium on the Genetics of Industrial Microorganisms* In Press. American Society for Microbiology.

HINNEN, A., HICKS, J.B. & FINK, G.R. (1978). Transformation of yeast. *Proceedings of the National Academy of Sciences* 75, 1929-1933.

HOPWOOD, D.A. (1978). The many faces of recombination. *3rd International Symposium on the Genetics of Industrial Microorganisms* In Press. American Society for Microbiology.

HOŠTÁLEK, Z., BLUMAUEROVÁ, M. & VANĚK, Z. (1974). Genetic problems of the biosynthesis of tetracyclines. *Advances in Biochemical Engineering* (Eds. T.K. Ghose, A. Fiechter and N. Blakeborough) 3, 13-67. Berlin and Heidelberg Germany and New York U.S.A.: Springer-Verlag.

JOHNSTON, J.R. (1975). Strain improvement and strain stability in filamentous fungi. *The Filamentous Fungi* (Eds. J.E. Smith and D.R. Berry) 1, 59-78, London U.K.: Arnold.

MACDONALD, K.D. (1972). Storage of conidia of *Penicillium chrysogenum* in liquid nitrogen. *Applied Microbiology* 23, 990-993.

MARTIN, J. (1978). Biochemical genetics of antibiotic biosynthetic pathways. *3rd International Symposium on the Genetics of Industrial Microorganisms* In Press. American Society for Microbiology.

MURRAY, K. (1976). Biochemical manipulation of genes. *Endeavour* 126, 129-133.

NAGAOKA, K. & DEMAIN, A.L. (1975). Mutational biosynthesis of a new antibiotic, streptomutin, by an idiotroph of *Streptomyces triseus. Microbial Genetics Bulletin* 39, 8.

NAKAYAMA, K., ARAKI, K., HAGINO, H., KASE, H. & YOSHIDA, H. (1976). Amino acid fermentations using regulatory mutants of *Corynebacterium glutamicum. 2nd International Symposium on the Genetics of Industrial Microorganisms* (Ed. K.D. Macdonald), pp. 437-449. London U.K. and New York U.S.A.: Academic Press.

PEBERDY, J.F. (1978). New approaches to gene transfer in fungi. *3rd International Symposium on the Genetics of Industrial Microorganisms* In Press. American Society for Microbiology.

PREER, J.R. (1971). Extrachromosomal inheritance : hereditary symbionts, mitochondria, chloroplasts. *Annual Reviews of Genetics* 5, 361-406.

RIGHELATO, R. (1976). Selection of strains of *Penicillium chrysogenum* with reduced penicillin yields in continuous culture. *Journal of Applied Chemistry and Biotechnology* 26, 153-159.
ROPER, J.A. (1973). Mitotic recombination and mitotic non-conformity in fungi. *1st International Symposium on the Genetics of Industrial Microorganisms* (Eds. Z. Vanek, Z. Hostalek and J. Cudlin), pp. 81-88. Amsterdam Holland, London U.K. and New York U.S.A.: Elsevier.
SERMONTI, G. (1969). *Genetics of antibiotic-producing micro-organisms*. London U.K., New York U.S.A., Sydney Australia and Toronto Canada: Wiley-Interscience.

FERMENTER DESIGN AND FUNGAL GROWTH

G.L. Solomons

RHM Research Limited
The Lord Rank Research Centre,
Lincoln Road, High Wycombe, Bucks. HP12 3QR

INTRODUCTION

A fermenter vessel provides a culture of an organism with its physical environment, and the interaction of this environment with the organism and its nutrient supply is often complex and little understood.

Paradoxically, many micro-fungi will produce biomass in almost any type of fermenter vessel, but to grow the same organism under well defined conditions is extremely difficult and rarely accomplished. This situation is brought about because many fungi will tolerate and continue to grow under conditions of low oxygen tension, local substrate limitation and local extremes of pH etc. As well as changes in metabolism, fungi will often respond to these adverse conditions by change of morphology, such as pellet formation or other alterations of mycelial form. Therefore the organism will often grow, albeit slowly, but the experimenter is frequently unaware of the true rate controlling condition in his experiment. Clearly, the organism and its nutrient supply play a major role in this situation, but here we will concentrate on the role of the fermenter vessel.

It is clear that there is no single fermenter configuration which is ideal for all purposes. Fermenters, as other equipment, call for different designs for different duties. Nevertheless, certain basic considerations have always to be taken into account in any good design, such as prevention of blockage of ports or inlet and outlet tubes, correct positioning of probes etc.

For short term experiments, up to 48-72 h duration, and where low cell concentrations can be used, 250-500 mg/ℓ dry weight, simple aerated (sparged) equipment is usually suitable.

Immediately more ambitious experiments are attempted, the limitations of this type of equipment become apparent. For many types of industrial research the need for high concentration of cells and metabolites means that the correct type of equipment must be available from the outset. The requirements for scale-up information are amongst the most demanding that laboratory investigations have to satisfy, although many investigators believe that this information is rarely adequate to explain behaviour on a larger scale.

It was the cultivation of *Penicillium chrysogenum* (Thom) for penicillin production that stimulated, in the late 1940's and early 1950's, the study of fermenter design. In 1956, three major U.S. pharmaceutical producers simultaneously published the designs of their small scale fermentation equipment (Finn & Sfat, 1956). The CSTR (continuously stirred tank reactor), fully baffled with vertical shaft driven turbine impellers and air spargeing proved to be the most common design and has, by and large, remained ever since the most used and useful design.

Attempts at other stirred vessel arrangements have been described, perhaps the best known being that of Means, Savage, Reusser & Koepsell (1962) which attempted to overcome a major problem of designing vessels for use with filamentous organisms, the blockage of process lines and overgrowth of the head space, by using a horizontal shaft equipped with a large number of simple flat blades. Although the vessel apparently worked satisfactorily, it has not been widely copied, probably because of its complexity and poor mechanical scale-up characteristics.

The main disadvantage of the CSTR was then considered to be the infection risk of the stirrer gland, especially prior to the widespread adoption of mechanical seals and, to overcome this drawback, magnetic couplings which do not require vessel penetration have been used (Cameron & Godfrey, 1974). These designs, whilst perhaps useful for small scale vessels (and particularly useful for pathogenic organisms) cannot be scaled up to any substantial size. Another approach to the problem of seal design is to entirely eliminate mechanical agitation and rely on air or oxygen to provide both aeration and agitation. Steel, Lentz & Martin (1955) adopted this system for a citric acid fermentation and more modern tower fermenters have been described by Greenshields & Smith (1974). It has been common knowledge that one major antibiotic producer, the Pfizer Company, has always used non-agitated vessels for a whole range of fermentation processes.

Current interest in non-agitated pressure-cycle loop fermenters, of great size and designed especially for single-cell protein production (Gibson, Roesler, Smith & Maslen,

1974) has undoubtedly re-stimulated interest in their potential application to multicellular organism fermentation. The re-cycle loop fermenter has been designed with an awareness of the importance of mixing time as a major parameter for the successful scale-up of fermentation processes. This parameter has emerged over the past few years as the most important new concept for fermenter design.

CHARACTERISTICS OF FUNGAL CULTURES

Fungi growing in submerged culture may exhibit three major characteristics: complex morphology by growing as apparent single cells or in pellet or in filamentous form: complex rheology as either 'solid' pellets in a non viscous Newtonian culture medium or as a non-Newtonian fluid; and growth on surfaces, either as a compact mass below the liquid level, on probes and baffles etc. or as surface growth, which may even be differentiated and form characteristic sporing structures on fermenter head plates above the liquid level.

Cell Morphology

In submerged culture, fungi have four possible growth forms, these are: small, discrete cells that closely resemble a yeast cell; small, compact hard pellets; larger, floccose pellets and true filamentous form (Fig. 1). The initiation of these different growth forms is a function of the organism itself, the method and amount of inoculum used, the culture medium employed and finally the physical conditions controlled by the fermenter, such as stirrer speed, shear, temperature, pH etc. The biochemical consequences of the particular growth form adopted are profound (Solomons, 1975), but suffice it to point out that the most immediate and perhaps critical effects are those on oxygen demand and supply. Besides the effect on growth rate, the morphological form plays an important role in determining the actual cell concentrations that can be reached by the organism. In our experience, 'single-cell' growth form allows high dry weights to be achieved, we have obtained a cell mass of 40 g/ℓ (dry weight) after 18 h incubation. Pellet growth of either form usually results in lower biomass formation, in the region of 10-20 g/ℓ. True filamentous growth can produce 50 or even 60 g/ℓ dry weight, although to achieve this concentration requires prolonged incubation time (72-96 h) and growth, due to oxygen limitation, is invariably linear.

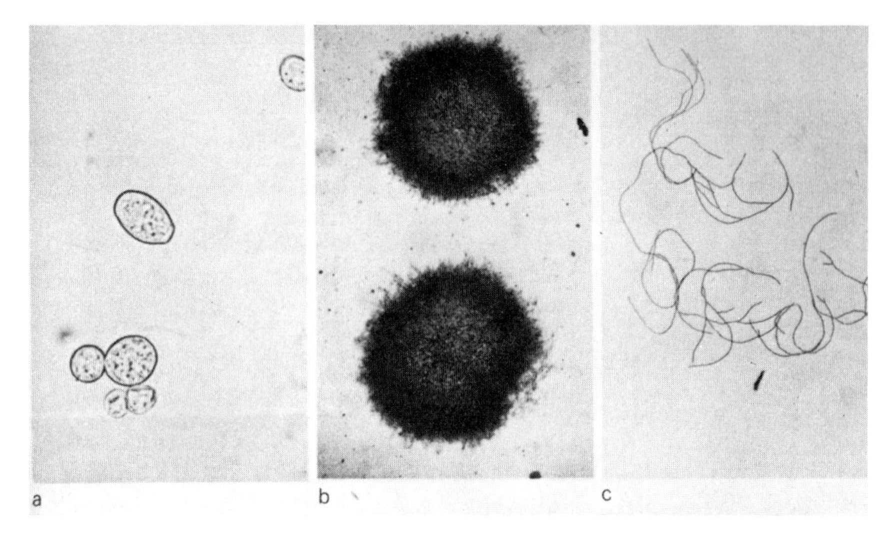

Fig. 1. Forms of fungal growth (a) single cell growth; (b) pelleted growth, and (c) filamentous growth (Solomons, 1975).

Culture Rheology

In order to achieve an understanding of the problems of designing fermenters for growing micro-fungi, it is necessary to appreciate the relationship between cell concentration and growth form on the one hand and the rheological properties that arise from these on the other hand. The rheological properties are important because the viscosity (both Newtonian and non-Newtonian) influences the turbulence of the system as expressed by the Reynolds Number (N_{Re}),

$\frac{D^2 N\rho}{\mu}$ or $\frac{D^2 N\rho}{\mu_a}$ where D = dia of impeller, N = rps of impeller, ρ= density of fluid, μ = viscosity of Newtonian fluid (poise), μ_a = apparent viscosity of non-Newtonian fluid measured at the same shear rate as that produced by the impeller.

At low N_{Re} the shear rate of the impeller = 10 N (Calderbank & Moo-Young, 1959) but at higher N_{Re}, shear values as high as 100 N are reached (Van't Riet & Smith, 1975). Turbulence is important in turn because of the role it plays in heat and mass transfer. It can be stated that the lower the turbulence (N_{Re}) of a fluid, the poorer will be the heat and mass transfer rates.

It therefore follows that increase in viscosity adversely

affects oxygen transfer and we have the paradox that increase in mycelium concentration leads to higher viscosity, which lowers the turbulence and therefore the oxygen supply, just as this is required to increase to support the greater mycelial biomass.

The universal and most important constituent of culture medium is water and it usually constitutes around 90% of most culture medium. Water is a Newtonian fluid (Fig. 2), that is, one in which there is a linear relationship between shear stress and shear rate (velocity gradient), (Eq. 1.):

$$\tau = \mu \, (dv/dy) \qquad \text{Eq.1.}$$

where τ = shear stress (dynes/cm^2)
μ = viscosity (poise)
dv/dy = velocity gradient (shear rate, sec^{-1})

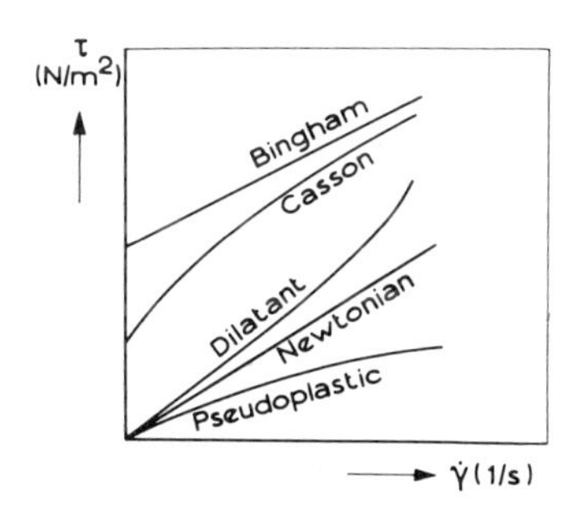

Fig. 2. Fluid rheology characteristics (Metz, 1976).

Most non-polymeric materials when dissolved in water increase the viscosity but allow it to retain its Newtonian characteristics. If we take as an example, a culture medium with 30 g/ℓ of glucose plus inorganic salts etc., the viscosity will increase only slightly from say 1.0 centipoise (water) to 1.5 cP. Even this increase is nullified by the temperature of fermentations, usually being in the range 30-55°C since the viscosity of water is 1 cP at 20°C, falling to 0.8 and 0.7 cP at 30 and 35°C respectively.

If we now grow in this Newtonian culture medium, small, regular organisms such as non-filamentous yeasts, most bacteria or fungi growing in the form of small cells, the viscosity characteristics of the culture retain a Newtonian characteristic, even though the value of 1 cP will rise in relationship to the actual size and concentration of the organisms present, according to the relationship

$$\mu_s = \mu_L \, (1 + f(\mathcal{J}, \phi)) \qquad \text{Eq.2.}$$

where μ_s = viscosity of the culture

μ_L = viscosity of the culture medium (water)

$f(J,\phi)$ = function of ratio, length to diameter J, of particles and the fractional volume ϕ occupied by the organisms.

In the case of dilute suspensions of spheres, the Einstein equation (Eq. 3) shows that the function $f\ (J,\phi)$ equals $2.5\ \phi$. At higher concentrations of suspended particles, Vand (1948) has extended the relationship to include particle-particle interaction, (Eq.4).

$$\mu_s = \mu_L\ (1 + 2.5\ \phi) \qquad \text{Eq.3.}$$

$$\mu_s = \mu_L\ (1 + 2.5\ \phi + 7.25\ \phi^2) \qquad \text{Eq.4.}$$

Deindoerfer & West (1960), reporting the work of Eirich, Bunzle & Margaretha (1936) show that with up to 3% suspended cells of yeast and *Lycoperdon* spores, the relative viscosity μ_s/μ_L rises to approximately 1.08 and even at 10% concentration, only rises to around 1.4. For practical purposes then, the growth of single cell organisms has little effect on the overall viscosities of the cultures and the values for their viscosities are not appreciably greater than that of water.

The use of polymeric materials in culture medium, e.g. starch and the production of polymeric materials by organisms e.g. polysaccharides, considerably alter the situation. Fortunately, in the former case, the higher initial viscosity coincides with minimum total oxygen demand by the culture and usually the organism produces amylolytic enzymes which reduce the viscosity of the starch as the organism grows and hence increases its total oxygen requirement. The production of polymers increases the viscosity of the culture, presenting even more severe problems of mixing and oxygen transfer than that caused by high concentrations of filamentous fungi (Lawson, 1976; Hubbard & Williams, 1977).

Pelletal growth of fungi usually provides a "background" of culture medium which is Newtonian in character and this has the property of low viscosity. This applies particularly to small, smooth hard pellets and to low concentrations of floccose pellets.

Clearly in the case of the latter, at some point the characteristics begin to merge with those of a true filamentous type of growth.

When fungi grow in filamentous mycelial form, the hyphae intertwine and produce a three dimensional network which has a structural rigidity (Deindoerfer & West, 1960). This structural rigidity is a most important concept, since although the mycelium only forms a minor weight component of the culture broth, it imparts *all* of the viscosity characteristics, since the "background" culture medium is usually of the same viscosity as water. This situation is unlike the rheological properties possessed by polymers in solution, and it may well be that it is not appropriate to use viscous non-Newtonian solutions as rheological models for mycelial networks.

The *gross* rheological properties of some polymer and filamentous cultures of fungi are described as either plastic or pseudoplastic (Fig. 2). Non-Newtonian fluids exhibit a non-linear relationship between shear rate and shear stress. Plastic or Bingham plastic fluids exhibit a yield stress, that is a force which is needed before they will flow and above which they behave in a manner similar to a Newtonian fluid. Pseudoplastic fluids exhibit decrease in shear stress with increasing shear rate and hence can be shear rate "thinned". This reduction in apparent viscosity with increasing shear rate approaches a limiting condition (Fig. 3). Examples of pseudoplastic fluids are long flexible molecules such as natural gums and synthetic thickeners such as carboxymethyl cellulose.

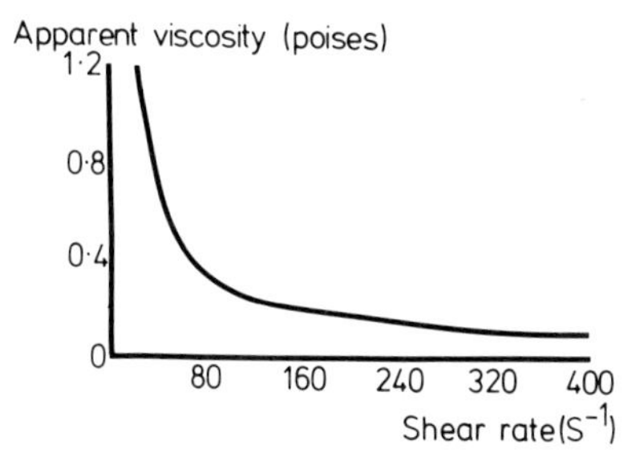

Fig. 3. Effect of shear rate on apparent viscosity of a filamentous mycelium culture broth (Solomons, 1975).

The rheological measurement of non-Newtonian fluids necessitates the use of rather complex equipment and because of the change of apparent viscosity with differing shear rates, a series of readings are required. Several makes of Couette (cup and bob) viscometers are available, but measurement at shear rates below 1 sec^{-1} requires special low shear rate measuring equipment. However, it is not just a question of the apparatus used, but perhaps of equal importance is the problem of representative sampling. Polymer solutions are usually uniform at the small sample level (i.e. 0.1 ml), but mould broths are often non-uniform in a much larger sample level (i.e. 10 ml). For this reason, we use samples of around 500 ml and have developed a measuring technique using a bob of a Couette viscometer rotating in an infinite field (Cheng & LeGrys, 1975). We have examined the method described by Roels, Van den Berg & Voncken (1974) who recommend the use of a disc-turbine impeller instead of a bob, but find that we can obtain equally reproducible results by our method using the power law relation

$\tau = K(\dot{\gamma})n$ where τ = shear stress (dyne/cm^2)

K = fluid consistency index, $\dot{\gamma}$ = shear rate (sec^{-1}),

n = exponent

The various components of a mycelial culture that produce its overall rheological properties, such as dry weight, hyphal length, branching, mycelium flexibility etc. have been described in great detail by Metz (1976) and to date this is undoubtably the single most important publication on the subject. From all of the evidence obtained, the rheological properties of mycelial cultures can be summarised in the description by Metz (1976) as: 1) The rheological properties of mycelial cultures can be well described by a Casson model. 2) The yield stress and Casson viscosity can be correlated with dry weight, the dimensionless length of the particles and branch frequency, expressed as lengths of the hyphal growth unit. Of these factors, mycelium concentration is by far the most important. 3) The flexibility of hyphae can have a considerable influence on the rheological behaviour of mycelial cultures. 4) Energy input, expressed as stirrer power has little effect on hyphal length. 5) A dynamic equilibrium between flocculation and deflocculation exisits and in cultures of high dry weight, this reaction is very rapid (<< 1 sec) and it is this basic mechanism which underlays the rheological behaviour of mycelial cultures.

The consequences of the rheological properties are of paramount importance for the design of satisfactory fermenters because they show that: 1) Essentially all the oxygen transfer in the fermenter takes place within the impeller region. 2) After the flocs of mycelium leave the impeller region, they tend to behave as discrete particles with little mass transfer between them. These observations lead to the conclusion that fermenter mixing time is one of the most important, perhaps the most important parameter of design, because after leaving the impeller region, no fresh oxygen or other nutrients are available to the cells until they receive fresh "charges" on their return to the region of high shear and low viscosity near the impeller.

THE CONTROL OF FUNGAL CULTURE BROTH RHEOLOGY

The two factors which determine the rheological properties of a fungal culture broth are: a) cell concentration and b) cell morphology.

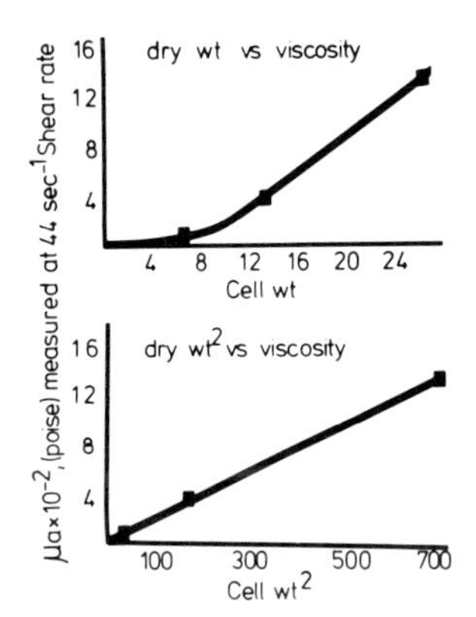

Fig. 4. Mycelium dry weight versus viscosity (Randall & Solomons, 1977).

Cell concentration is a simple parameter to control, but decrease in cell concentration can lead to loss of product yield. Japanese workers have suggested that the addition of about 10% of water at the end of a batch culture reduces viscosity sufficiently to enable aeration to be affected (Taguchi, 1971). From Fig. 4 it can be seen that the addition of 10% of water would reduce by 14.5% the apparent viscosity of a culture broth, initially at 25 g/ℓ dry weight. This technique has other obvious disadvantages. Therefore, more attention has been paid to alteration of cell morphology. Two methods have been examined, the first being the alteration from filamentous growth to pellet growth and whilst this method favourably alters the overall culture rheology, it does impose other limitations, especially of nutrient diffusion into the pellets (Bhavaraju & Blanch, 1975). The second is to change the cell morphology by changes in cultivation conditions or to try to break down the hyphal structures by mechanical force by means of an agitator.

The importance of culture pH on the morphological form of *Penicillium chrysogenum* grown in continuous culture was described by Pirt & Callow (1959), who found that the hyphal length decreased with increase in pH above 6.0 so that they achieved a minimum value for hyphal length at pH 7.0-7.4 and above. Without alteration of culture pH, a number of fungi when grown in continuous culture will spontaneously alter their morphological form to produce shorter, more highly branched mycelium, which considerably reduces culture viscosity and thus reduces the problem of adequate aeration. Such a change was described by Forss, Gadd, Lundell & Williamson (1974), who observed this occurring with a strain of *Paecilomyces varioti* (Bainier) after about one week in continuous culture. We have observed similar changes in a strain of *Fusarium graminearum* (Schwabe) after 4-6 weeks in continuous culture, but we observed no such change with a strain of *P. chrysogenum-notatum*. Because there are apparently differences in the time required for the appearance of these morphologically altered strains, perhaps our *Penicillium* would have produced these altered forms if the culture had been extended in time to beyond 4-6 weeks.

The alteration by mechanical means of hyphal morphology in a strain of *P. chrysogenum* has been described by Dion, Carilli, Sermonti & Chain (1954), but with another strain of *P. chrysogenum* Kossen & Metz (1976) have shown that power input by an impeller has almost no influence on the length of the hyphae. They concluded that shear (impeller rpm) does not provide a practical means of decreasing the viscosity of mould suspensions. Our experiments with a strain of *F. graminearum* (Randall & Solomons, 1977) confirm the findings of Kossen & Metz (1976). Continuous culture experiments using different impeller sizes and diameters in combination with different stirring speeds showed that the mean hyphal length was virtually unaffected over a wide range of experimental conditions. The experiments of both Kossen & Metz (1976) and Randall & Solomons (1977) were carried out using cells growing in continuous culture. It is however possible to show, by simple stirring experiments, that agitation can have a profound effect on culture rheology. Randall & Solomons (1977) have also shown that a sample of culture grown in a $1.3m^3$ continuous culture vessel when stirred in a 3 l laboratory vessel with temperature and pH control, did appreciably alter its viscosity, Figs. 5 & 6. The cause of this reduction in viscosity is not yet known, it may reflect a breakdage of the hyphae, or it may be a disentanglement of the hyphal network leading to a reduction in flow resistance.

Another method that has been employed to alter the physical properties of fungal culture broths is the addition of water soluble polymers (Moo-Young, Horose & Geiger, 1969; Elmayergi, 1975). Polymers and fibres are known to reduce drag in fluids (Ellis, 1970) while Randall & Solomons (1977) have unsuccessfully attempted to lengthen the hyphae of fungi grown in continuous culture by the addition of carboxypolymethylene.

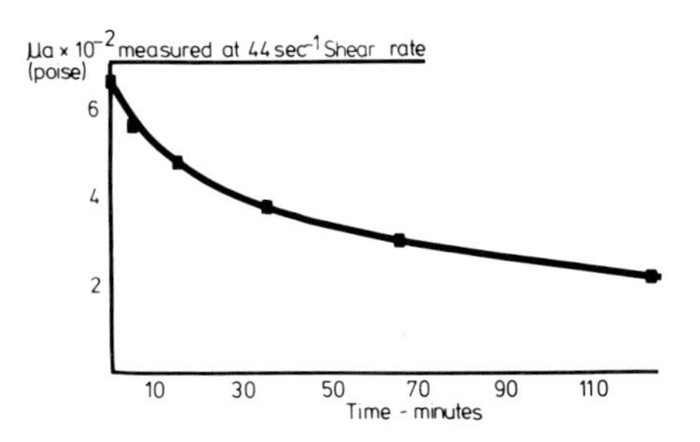

Fig. 5. Viscosity of fungal culture versus time of stirring (Randall & Solomons, 1977). Culture grown in $1.3m^3$ fermenter and sample transferred into a 3 l laboratory fermenter equipped with a 7.5 cm turbine and stirred at 1000 r.p.m.

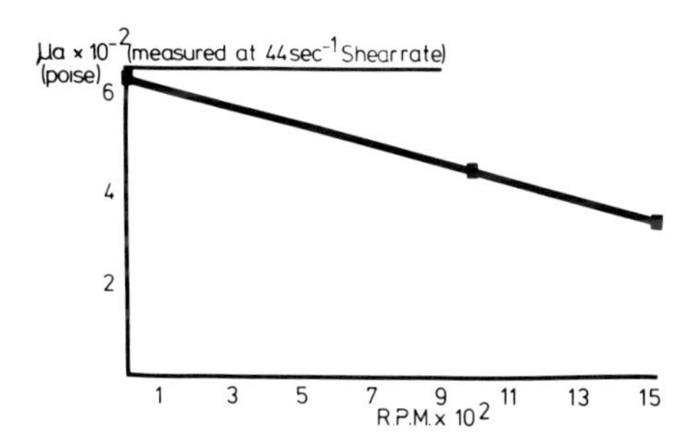

Fig. 6. Viscosity of fungal culture versus stirring at different r.p.m. for 1 h (Randall & Solomons, 1977). Culture grown in $1.3m^3$ fermenter and sample transferred into a 3 l laboratory fermenter.

VESSEL CONFIGURATION

One of the earliest fermenters designed for growing microfungi in submerged culture, which was used at the Northern Regional Research Laboratory, Peoria, Illinois, U.S.A., is reviewed by Prescott & Dunn (1959) who also describe the extensive use made of this equipment. The horizontal rotating cylinder design was built in two sizes, 3.2 and 540 l for laboratory and pilot plant use respectively. Its major disadvantages were that to accommodate air inlet and outlet pipes, the vessel could only be run one-third full and its oxygen transfer capacity (not measured at the time) must have been very limited.

With the advent of the submerged culture production of penicillin, vertical stirrer tanks were used, an early example being the mild steel tanks described by Stefaniak, Gailey, Brown & Johnson (1946). Subsequent equipment was mostly very similar and was based on stirred vessels with a diameter (D) : height (H) ratio of around 1:1 to 1:2 and equipped with turbine type impellers (Fig.7). Large scale equipment was also based upon the vertical stirred tank with a D/H ratio of 1:2 to 1:3 with the additional feature that multi-impellers were mounted on the centrally placed stirrer shaft. On the larger scale, around 1 HP/100 gals of culture was normally provided for agitation. A typical installation was illustrated in an advertisement by Lightnin Mixers Ltd., who showed a 15 ft dia 45,000 US gal vessel equipped with 3,6 ft dia disc-turbines spaced along a 40 ft agitator shaft and driven by a 450 HP motor. All of these designs used baffled tanks and were equipped with spargers for introducing air into the bottom of the vessel. Spargers were single open pipes, rings, or more occasionally sintered. The latter tend to block with mycelium in a short time and simple open pipes have become the more favoured type.

From quite early on, concern was being expressed about the availability of oxygen to the cells (Bartholomew, Karow, Sfat & Wilhelm, 1950; Rolinson, 1952). Then Finn (1954) published his review which focussed attention upon the problems of aeration and agitation. This classic work still repays reading many years after its publication.

In seeking to overcome the oxygen transfer rate limitation in fungal culture broths, Chain and his collaborators described the use of mechanically stirred vessels that could use vortex aeration to replace subsurface spargeing. The problems of the system were that the vessel contents had to be reduced in volume because of the vortex and although the system was scaled up to 3000 l size, it was not considered practical for commercial sized vessels (Chain, Paladino, Ugolini, Callow & van der Sluis, 1954; Paladino, Ugolini & Chain, 1954).

As an alternative to the stirred fermenter, Steel *et al.* (1955) proposed the use of an aerated tower fermenter and described its application to citric acid fermentation. Their fermenters were of only 2.5 and 36 l capacity with a D/H ratio of 1:2.5 but since then other workers have used larger tower fermenters (Greenshields & Smith, 1974). Many of the fermentations described in tower fermenters have used pellet type growth to avoid the complexities associated with oxygen transfer into the culture, although the problem remains of oxygen and nutrient transfer into the mycelium of the pellet.

Undoubtably, such fermenter designs are simpler and mechanically more reliable than stirrer vessels, but it is true to say that most industrial fermenters are based on the stirred tank design. Oxygen transfer is often equated with total power input into the vessel, agitation by impeller is around 90% efficient in terms of electrical energy from the motor into power dissipated by the impellers. Air compression is only around 40% efficient so that it is more than twice as effective in electrical energy terms to agitate a culture by impeller stirring, than it is to dissipate the same power by aeration.

IMPELLER DESIGN

The purpose of providing an agitator or impeller (Fig. 7) in a tank is to obtain mixing of the vessel contents and to achieve heat and mass transfer. Heat transfer is required for efficient heating and cooling for temperature control purposes and mass transfer is required to provide a sufficient supply of dissolved and other nutrients to the cell and to remove carbon dioxide from the culture. It is still a widespread misconception that power uptake into a stirred tank is determined by the prime mover, usually an electric motor. In fact the power uptake in a stirred tank is largely determined by the impeller system employed. To stir a given fluid at a set speed with an impeller requires a certain amount of power. Failure to provide that power, i.e. too small a motor, and the impeller speed falls, but power in excess of the impeller demand is simply unused for the purposes of agitation.

Early fermenter designs rather assumed that all impeller systems were more or less equal in their effectiveness and hence a wide variety of impellers were used. Many of these designs were based on, or were adaptations of, marine propellers; the Pfaulder vessel used by Gordon *et al.* (1947) is an example of this type. Another design favoured largely because of ease of fabrication, is the flat paddle. The two outstanding publications by Cooper, Fernstrom & Miller (1944) on the use of sulphite oxidation for measuring oxygen transfer rates and Rushton, Costich & Everitt (1950) on power uptake by impellers, led to the widespread adoption of the 6, flat-blade disc turbine impeller. This design has many desirable features: it is easy to fabricate, the design is favourable for scaling-up and if correctly designed and installed it provides excellent mass and heat transfer capability. Over the past 25 years or so, a vast body of chemical engineering literature has appeared which has lead to an understanding of the role of the disc turbine in fermentation systems, (see for example Calderbank, 1967; Miura, 1976). Disc turbines do have a high power demand and attempts have been made to find satisfactory alternative designs. The rod-impeller of Steel & Maxon (1966) attempted to disperse the power dissipation throughout a greater volume of the fermenter than does a turbine and these authors did report a

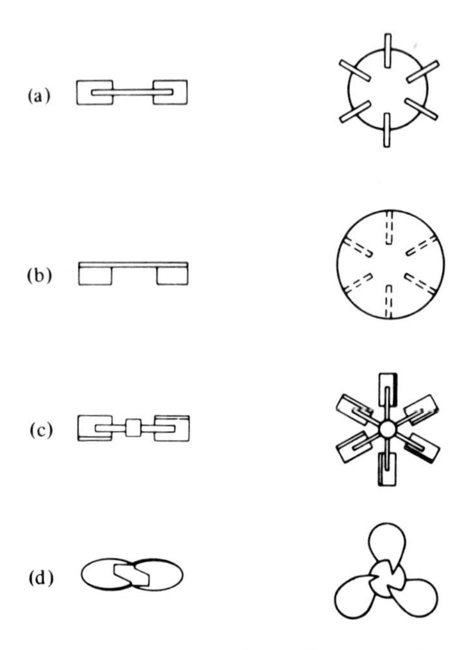

Fig. 7. Types of impellers: (a) disc turbine; (b) vaned disc; (c) open turbine, variable pitch; (d) marine propeller (Solomons, 1969).

considerable reduction in power input with no loss of product formation. A double cone impeller was used by Parker & Jackson (1964), which the authors describe as acting as a centrifugal pump, and low power uptakes and no loss of enzyme titre was claimed with an *Aspergillus* culture. We have used both of these designs in a 130 l scale fermenter and were unsuccessful with either impeller design in obtaining satisfactory growth rates with a *Penecillium* sp.

Until recently, the actual manner in which disc turbines effected oxygen transfer was not clear. It had been supposed for example that the blades "chopped up" air bubbles and this created smaller bubbles with larger surface areas etc. However, since the early 1970's a series of publications from the University of Delft have appeared which has done much to clarify the way in which disc turbines effect oxygen transfer (Van't Riet & Smith, 1973; Van't Riet & Smith, 1975; Van't Riet, Boom & Smith, 1976; Smith, Van't Riet & Middleton, 1977). From their work, it is now apparent that each blade of a disc turbine trains a pair of curved vortices, one above and the other below the disc, and all the gas supplied to a fermenter passes through these "gas-traps" with bubbles breaking off at the end of the vortices.

These workers have also shown that within limits, the larger number of blades on the turbine, the larger is the amount of gas that the impeller can effectively deal with. It had been known for a long time that different designs of impeller could effectively only disperse gas up to a critical superficial velocity (Sv), i.e. velocity of gas flow/unit area measured across the diameter of the fermenter vessel. As vessels increase in size, their cross-sectional area/volume ratio decreases, but since the demand for oxygen and hence air supply is usually independent of scale (ignoring differences in mass-transfer efficiency), the Sv usually increases with scale-up. Conventional six bladed disc- turbine impellers can effectively disperse gas up to a Sv of around 400 ft/h, compared with 70 ft/h for a marine impeller (Finn, 1954). The Delft workers have shown that by increasing the number of blades on the disc from the usually 6 to 12 or even 18, the amount of gas that can be effectively handled can be increased threefold.

It is generally accepted that most (80%+) of the oxygen transfer within a fermenter, particularly in a fermenter growing filamentous fungi, occurs within the region of the impeller envelope. The disc-turbine generates a flow pattern which is described as "radial", that is, it draws in fluid to its centre and discharges fluid as a jet from the end of the blades, thus setting up a re-circulation loop. The time that a particle of mycelium takes to re-cycle back into the impeller envelope, and hence into an area of high dissolved oxygen, is a most important parameter. The level of dissolved oxygen, even at saturation, is only a few parts per million so that if there is a high demand for oxygen by the cells, then the supply of fresh oxygen that the cell picks up as it passes through the impeller envelope may only be sufficient for the cells's respiration for a matter of a few seconds, typically between 10-60 seconds. If the culture is not to be limited by oxygen supply, then assuming that the oxygen transfer rate is not limiting, the mixing time of the impeller must be of a similar order, i.e. 10-60 seconds. Much attention is now being paid to fermenter mixing time, which increases as vessel volume increases according to the empirical formula $T_{mix} \alpha (V)^{0.3}$ (Eirsele, 1976, 1978) and it can be readily appreciated that for large vessels this parameter may govern the productivity of the system. Romantschuk & Lehomäki (1978) have described how, in a 360m^3 fermenter, a considerable volume of 'dead-space' caused by poor mixing decreased by 40% the expected biomass productivity. Mixing time (T_{mix}) is defined as the time in seconds taken for a pulsed disturbance to the system to be "smoothed" and Fig. 8 illustrates how this can be measured by pH shifts

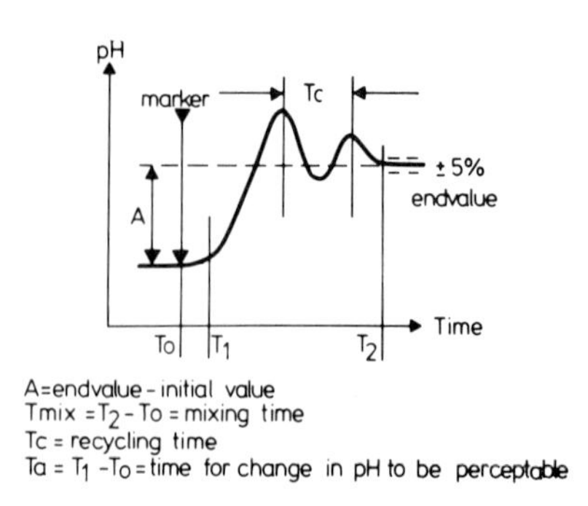

Fig. 8. Mixing time after pH adjustment by addition of acid or alkali (Einsele, 1976).

(Einsele, 1976, 1978). In this figure, it can be seen that the re-circulation loop time is approximately 30% of the time for total mixing and T_{mix} times for non-viscous culture broths are given as 5, 10, 20, 45 and 100 s for 0.01, 0.1, 1, 10 and 100m ^{3} vessels respectively. Mixing times for viscous fungal culture broths increase by up to an order of magnitude depending upon the viscosity, but it may be that re-circulation time, provided the 're-circulation loops' affect all of the vessel contents, is a more practical value to use since as long as the culture receives sufficient oxygen to remain above its C_{crit} value, it will not be oxygen limited. Two factors influence the mixing time of a fermenter; (a) the rheological properties of the culture and (b) the impeller system and fermenter configuration. It was shown by Metzner, *et al.* (1961) that dependent upon vessel configuration, a minimum degree of turbulence (N_{Re}) was necessary to permit mixing to occur and below a critical value, complete mixing could never be achieved. Therefore, the more the viscosity of the culture rises, by cell dry weight or lengthened hyphae or both, the more difficult it becomes to provide adequate mixing. Mixing time is also greatly influenced by the number and type of impellers (Metzner *et al.*, 1961); multi-impeller systems generally providing an improvement in mixing performance, i.e. a decrease in mixing time, especially when the vessel H/D ratio is increased above 1. Disc turbines by their flow pattern established a quite rigorous "sphere of influence" and we have found, in studying multi-impeller stirring, that there is a surprisingly slow transfer of fluid from one disc turbine impeller to the one above (LeGrys, 1977).

The first attempts to provide additional bulk flow mixing in a fermenter, was suggested by Jensen, Schultz & Shu (1966) who used a disc turbine combined with a marine type axial flow propeller. The disc turbine was intended to provide the bulk of the mass transfer whilst the marine propeller generated most of the bulk flow and therefore the mixing. This concept, whilst correct in theory, is not of much practical value in the laboratory vessels and our investigations suggest that the bulk mixing provided by using disc-turbine impellers is probably adequate for vessels with a capacity of up 50 m^3. Above this size, bulk mixing predominates in the effective design of "high performance" fermenters, since with adequate mixing more efficient use is made of the mass transfer capabilities of the fermenter and heat transfer is improved by increasing the velocity of culture over the heat exchanger surface.

MODERN TRENDS IN FERMENTER DESIGN

Laboratory sized fermenters of, say, 5-20 l capacity can be constructed around glass vessels with metal end-plates (Solomons, 1969). Mechanical agitation with a central stirring shaft mounting a single or double disc-turbine (d/D = 0.33-0.5) is totally adequate. Vortex aeration has been favoured by Rowley & Bull (1973) (Fig. 9), but we continue to favour a conventional baffled sparged fermenter design. When operating as a continuous culture vessel, it is advisable to run the vessel completely full so that no surface growth can occur because this invariably causes blockage of exit lines, etc. Comparatively little can be done to prevent growth occurring on the submerged surfaces of the fermenter such as stirrer shaft, baffles, pH probes, etc. Rowley & Bull (1973) have advocated the use of Teflon tubing whenever possible to provide the minimum hold for mycelium and we have treated all the internal surfaces of equipment with a silicone finish (Siliclad, Clay Adams Co., New Jersey, U.S.A.), but these treatments make rather marginal differences. The best methods of avoiding "wall growth" is to choose a strain which has a greatly reduced tendency to grow in this inconvenient fashion. As an example, we changed from a strain of *P. chrysogenum-notatum*, which in a few days, overgrew the internal surfaces of the fermenter, to a strain of *F. graminearum*, which allows us to operate for many weeks with only minor wall growth. We find that even in laboratory fermenters, pH probes are comparatively little affected by accreted growth provided they are placed sufficiently close to the impeller to be washed by its discharge jet. Dissolved oxygen electrodes are more of a problem and we usually arrange for a small jet of condensate/steam to impinge upon the Teflon surface of our Borkowski & Johnson (1967) type electrodes. By judiciously "washing" on

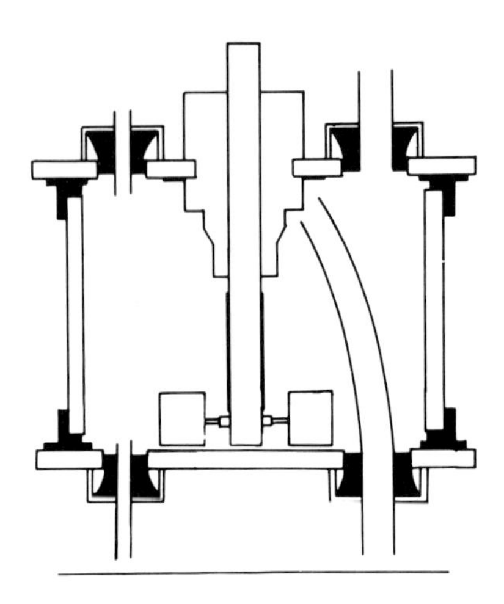

Fig. 9. A stirred tank reactor vessel for the continuous flow culture of moulds (Rowley & Bull, 1978).

a daily basis, we can keep the electrode sufficiently free of mould accretion to obtain sensible readings.

In running a small continuous culture vessel with filamentous fungi undoubtedly the greatest problem is that of 'feed back' of mycelium. In this situation, the culture vessel has a higher biomass concentration than does the exit stream. This complicates the kinetics of the system and various steps have been suggested to overcome the difficulties, Righelato & Pirt (1967) used a pinch valve which intermittently allows a forcible discharge of the excess volume in the vessel. An intermittent increase in stirrer speed has been recommended by Brunner & Röhr (1972) whilst Rowley & Bull (1973) favour the design illustrated in Fig. 9, which does not disturb the dilution rate, as do the other two methods proposed. We find that operation at high dilution rates >0.2 h^{-1} tends to cure the problem, but this solution is rarely available. Even at lower dilution rates, operating the fermenter at low cell dry weights, in the range 5-8 g/ℓ tends to prevent 'feed back'. On a 1.3m^3 pilot plant fermenter we have never experienced a problem of mycelium 'feed back'.

Sampling tubes should be placed so that the inlet to the tubing is near the impeller, thus ensuring that the sample, when taken, is as "mixed" as possible: additional tubes should also have their outlets near the impeller. When air spargers are used, we find that the simple hole type is the most suitable as it shows the least tendency to block. Tubing which is expected to carry mycelial culture broth should be correctly sized, we use a minimum of 10-12.5mm dia tubing since smaller sizes often lead to blockages.

As the volume of fermenter increases, the major problem which arises is lack of fluid bulk mixing, and from this deficiency stems a series of consequences such as poor overall oxygen transfer and poor heat transfer which causes changes in cell metabolism in turn affects the course of the fermentation. In a recent review (Katinger, 1977), emphasis has been given to 'recycle' fermenters, which provide greater bulk flow than do the traditional designs of stirred fermenters (Fig. 10). In design a fermenter with high mass transfer rates and mixing efficiency, Karrer, Einsele & Fiechter (1976) used the internal re-cycle loop principle and gave much emphasis to adequate baffling to prevent energy dissipation in tangential flow resistances. Kipke (1975) has described a similar design, but has also described a 100m^3 vessel, 12.4m high and equipped with a single stirrer shaft, carrying a number (up to six or more) of proprietary impellers (Mehrstufen-Impuls-Gegenstromrührer MIG) to provide the necessary agitation. It is claimed that these impellers are more efficient than conventional turbines, but they are probably most suited to yeast or bacterial fermentations rather than for filamentous fungi with their highly viscous culture broths. Because we have now achieved a better understanding of the mechanism of action of turbine impellers, we may expect substantial advances in improved designs. As Day (1976) has stated "Our studies also form a sound basis for the innovation of better devices for gas-liquid agitation and reaction. Ideally, these new devices will disperse gas without cavity formation while being based on sound principles so that scale-up will no longer be a problem. An additional target is a substantial improvement in energy efficiency, especially important in the case of larger reactor vessels".

Undoubtedly, the most spectacular advance in fermenter design has been the pressure re-cycle fermenter that has been devised by I.C.I. Ltd. for the production of SCP from methanol (Gibson, *et al.*, 1974). Their original external loop design has been considerably modified (Andrews, 1978) and, as installed, will now consist of a 42m high section with internal draft tube, topped by an 18m high expansion/gas detrainment section. The total volume is a. 5000m^3, the gas/liquid volume a. 3000m^3 and the liquid volume a. 2500m^3. The riser is fitted at intervals of a. 2m, with baffle plates with a fractional free opening of a. 0.3 in order to redistribute the air which is supplied at a rate of a. 1.6 VVM. On this truly collossal scale, bulk mixing times are long and the organism must experience a wide range of conditions within the fermenter. This vessel is designed for low viscosity culture broths and it would probably be unsuitable

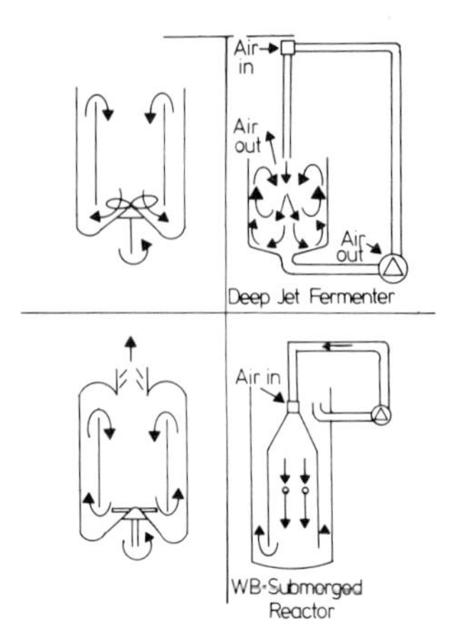

Fig. 10. New concepts of fermenter design (Katinger, 1977).

for filamentous fungal cultures, although it may be possible to use it with small pellet type growth. The reason for its unsuitability for viscous culture broths, is that above viscosities of 70cP bubble coalescence is enhanced by a factor of about tenfold, leading to the formation of spherical cap bubbles and a corresponding reduction in oxygen transfer (Calderbank, Moo-Young & Bibby, 1964).

We have approached the problem of the design of large, production scale fermenter vessels for use with filamentous fungi and hence viscous culture broths, by separating the function of bulk flow and mass transfer (LeGrys & Solomons, 1977). To this end we have used two impellers, each of optimal design for its function, and operating at its optimal conditions. This has lead to the overall design shown in Fig. 11 in which we have two stirrer shafts, one top and the other bottom entry, the shafts are very short and operate at different speeds. The bottom stirrer is a disc-turbine fitted with up to 18 blades to enable it to handle large volumetric air flow rates (Bruijn, Van't Riet & Smith, 1974; Smith & Van't Riet, 1976). This stirrer requires a large electric motor to drive it, but this can be mounted on the ground below the fermenter. The top stirrer, which takes less than 10% of the total power input, drives a specially designed low speed impeller which provides a flow velocity of around 1-1.5 m/s. This impeller, a Sabre-blade, is mounted in a baffled draught-tube and because the induced downwards velocity is higher than the free air bubble rise velocity, air is introduced by a sparger mounted just below the Sabre-blade, and is then driven down to the disc turbine for effective mass transfer. Introducing the air near the top of the fermenter results in very substantial savings in

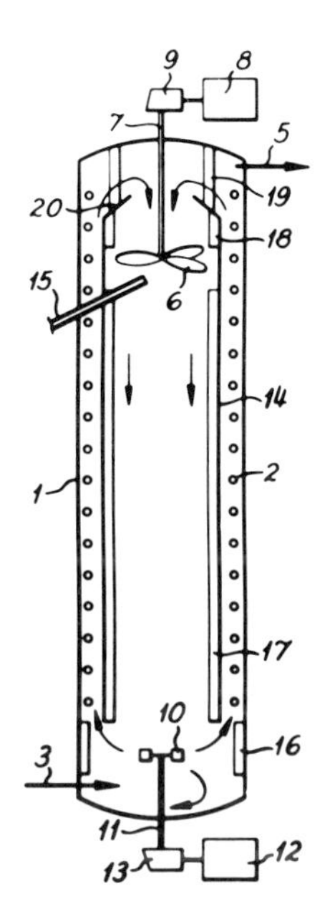

Fig. 11. Large scale industrial fermenter designed for short mixing time (LeGrys & Solomons, 1977).

energy for compression. The draught tube is also a heat exchanger and greatly contributes to providing the necessary heat exchange surface required in large fermenters. In this type of vessel, the Sabre-impeller in the downcomer pumps one tank volume in 20-30s, thereby ensuring that the cells in the system return to the highly oxygenated part of the fermenter before they have totally depleted their oxygen supply. Culture and air return up the riser, in which are mounted additional cooling coils. At its top, the draught tube has an inverted funnel shape which substantially reduces the rise in velocity causing the air to detrain and leave the fermenter. Re-entrainment has to be minimised because re-cycled air, which is depleted of oxygen, reduces the efficiency of the system.

Because the design eliminates long stirrer shafts, very tall vessels can be constructed with diameters kept below 16 ft which is the maximum diameter for rail transport. It is advantageous to have fabrication carried out in a construction shop rather than on-site because of the additional expense of the latter.

We believe that large scale vessels can be built using our design and although they will not be able to rival in size the ICI design, they will be suitable for growing filamentous fungal cultures.

REFERENCES

ANDREWS, S.P.S. (1978). Paper presented at Symposium. "Trends in fermenter design", S.C.I., London, May 18th.

BARTHOLOMEW, W.H., KAROW, E.O., SFAT, M.R. & WILHELM, R.N. (1950). Oxygen transfer and agitation in submerged fermentations. Mass transfer of oxygen in submerged fermentation of *Streptomyces griseus*. *Industrial & Engineering Chemistry* 42, 1801-1809.

BHAVARAJU, S.M. & BLANCH, H.W. (1975). Mass transfer in mycelial pellets. *Journal of Fermentation Technology* 53, 413-415.

BORKOWSKI, J.D. & JOHNSON, M.J. (1967). Long-lived steam-sterilizable membrane probes for dissolved oxygen measurement. *Biotechnology and Bioengineering* 9, 635-639.

BRUIJN, W., Van't RIET, K. & SMITH, J.M. (1974). Power consumption with aerated Rushton turbines. *Transactions of Institution of Chemical Engineers* 52, 96-104

BRUNNER, H. & RÖHR, M. (1972). Novel system for improved control of filamentous microorganisms in continuous culture. *Applied Microbiology* 24, 521-523.

CALDERBANK, P. (1967). Mass transfer in fermentation equipment. In *Biochemical and Biological Engineering Science* Volume 1, pp 101-180 (Ed. Blakebrough, N.). Academic Press, London and New York.

CALDERBANK, P.H. & MOO-YOUNG, M.B. (1959). The prediction of power consumption in the agitation of non-Newtonian fluids. *Transactions of Institution of Chemical Engineers* 37, 26-33.

CALDERBANK, P.H., MOO-YOUNG, M.B. & BIBBY, R. (1964). Coalescence in bubble reactors and absorbers. *Proceedings of Third European Symposium Chemical Reactors in Engineering* Amsterdam, September 15th-17th.

CAMERON, J. & GODFREY, E.I. (1974). Magnetic drives. *Biotechnology and Bioengineering Symposium*, No. 4, 821-835.

CHAIN, E.B., PALADINO, S., UGOLINI, F., CALLOW, D.S. & van der SLUIS, J. (1954). A laboratory fermenter for vortex and sparger aeration. *Rend. Inst. Sup. Sanita* 17, English Edition 61-86.

CHENG, D.C.H. & LeGRYS, G.A. (1975). Some experience in the use of a co-axial cylinder viscometer on fermentation broths. Paper presented 2nd Annual Research Meeting of Institution of Chemical Engineers, Bradford, U.K. March 24th 25th 1975.

COOPER, C.M., FERNSTROM, G.A. & MILLER, S.A. (1944). Performance of agitated gas-liquid contactors. *Industrial and Engineering Chemistry* 36, 505-509.

DAY, R.L. (1976). Process Technology. *Chemistry and Industry* 7th February, 81-83.

DEINDOERFER, F.H. & WEST, J.M. (1960). Rheological properties of fermentation broths. *Advances in Applied Microbiology* 2, 265-273.

DION, W.M., CARILLI, A., SERMONTI, G. & CHAIN, E.B. (1954). The effect of mechanical agitation on the morphology of *Penicillium chrysogenum* Thom in stirred fermenters. *Rend. Ist. Sup. Sanita* 17, English Edition 187-205.

EINSELE, A. (1976). Charakterisierung von Bioreaktoren Durch Mischzeiten. *Chemische Rundschau* 29 (25), 53 & 55.

EINSELE, A. (1978). Scaling up bioreactors. *Process Biochemistry* 13 (7), 13-14.

EIRICH, F., BUNZL, M. & MARGARETHA, H. (1936). Viscosity of suspensions and solutions. Part 4. The viscosity of suspensions of spheres. *Kolloid Z.* 74, 275-285. viz. *C.A.* (1936) 30, 7418 (1).

ELLIS, H.D. (1970). Effects of shear treatment on drag-reducing polymer solutions and fibre suspensions. *Nature, London* 226, 352-353.

ELMAYERGI, H. (1975). Mechanisms of pellet formation of *Aspergillus niger* with an additive. *Journal of Fermentation Technology* 53, 722-729.

FINN, R.K. (1954). Agitation-aeration in the laboratory and in industry. *Bacteriological Reviews* 18, 254-274.

FINN, R.K. & SFAT, M.R. (1956). Improvements in fermentation equipment and design. *Industrial & Engineering Chemistry* 48, 2172-2215.

FORSS, K.G., GADD, G.O., LUNDELL, R.O. & WILLIAMSON, H.W. (1974). Process for the manufacture of protein-containing substances for fodder, foodstuffs and technical applications. U.S. Patent No. 3,809,614.

GIBSON, M.R., ROESLER, F.C., SMITH, S.R.L. & MASLEN, F.P. (1974). Fermentation method and fermenter. B.P. 1,353,008.

GORDON, J.J., GRENFELL, E., KNOWLES, E., LEGGE, B.J., McALLISTER, R.C.A. & WHITE, T. (1947). Methods of penicillin production in submerged culture on a pilot plant scale. *Journal of General Microbiology* 1, 187-202.

GREENSHIELDS, R.N. & SMITH, E.L. (1974). The tubular reactor in fermentation. *Process Biochemistry* 9, (3), 11-17, 28.

HUBBARD, D.W. & WILLIAMS, C.N. (1977). Continuous fermenters for polysaccharide production. *Process Biochemistry* 12 (10), 11-13.

JENSEN, A.L., SCHULTZ, J.S. & SHU, P. (1966). Scale-up of antibiotic fermentations by control of oxygen utilisation. *Biotechnology and Bioengineering* 8, 525-537.

KARRER,D., EINSELE, A. & FIECHTER, A. (1976). Reactor design for high mass transfer rates and mixing efficiency. *Fifth International Fermentation Symposium*, Berlin, Abstract 4.11. p.67.

KATINGER, H.W.D. (1977). New fermenter configurations. In *Biotechnology and Fungal Differentiation*. pp 137-154 (Eds. Meyrath, J. and Bu'Lock, J.D.). Academic Press Inc. New York and London. „
KIPKE, K.D. (1975). Ruhrtechnische probleme bei der auslegung von fermentern. *Chemie-Technik* 4, 383-388.
KOSSEN, N.W.F. & METZ, B. (1976). The influence of shear upon the morphology of moulds. *Fifth International Fermentation Symposium*, Berlin, Abstract 4.22. p. 78.
LAWSON, C.J. (1976). Microbial polysaccharides. *Chemistry and Industry* March 20th, 258-261.
LeGRYS, G.A. (1977). Unpublished results.
LeGRYS, G.A. & SOLOMONS, G.L. (1977). Patent application 23128.
MEANS, C.W., SAVAGE, G.M., REUSSER, F. & KOEPSELL, H.J. (1962). Design and operation of a pilot plant fermenter for the continuous propagation of filamentous microorganisms. *Biotechnology and Bioengineering* 4, 5-16.
METZ, B. (1976). From pulp to pellets. PhD Thesis, Delft University of Technology.
METZNER, A.B., FEEHS, R.H., RAMOS, H.L., OTTO, R.E. & TUTHILL, J.D. (1961). Agitation of viscous Newtonian and non-Newtonian fluids. *American Institution of Chemical Engineers Journal* 7, 3-9.
MOO-YOUNG, M.B., HIROSE, T. & GEIGER, K.H. (1969). The rheological effects of substrate-additives in fermentation yield. *Biotechnology and Bioengineering* 11, 725-730.
MIURA, Y. (1976). Transfer of oxygen and scale-up in submerged aerobic fermentation. *Advances in Biochemical Engineering* 4, 3-40.
PALADINO, S., UGOLINI, F. & CHAIN, E.B. (1954). Fermenters of 90 and 300 litre capacity for vortex and sparger aeration. *Rend. Ist. Sup. Sanita* 17, English Edition 87-120.
PARKER, K.M. & JACKSON, A.T. (1964). A scale-up problem in relation to enzyme production involving the agitation of mould mycelium in the preparation of fungal amylases. Paper presented to Institution of Chemical Engineers Meeting, Birmingham, U.K. July 17th.
PIRT, S.J. & CALLOW, D.S. (1959). Continuous flow culture of the filamentous mould *Penicillum chrysogenum* and the control of its morphology. *Nature, London* 184, 307-310.
PRESCOTT, S.C. & DUNN, C.G. (1959). Industrial Microbiology. McGraw-Hill Book Co., Inc., New York and London.
RANDALL, A.J. & SOLOMONS, G.L. (1977). Unpublished results.
RIGHELATO, R.C. & PIRT, S.J. (1967). Improved control of organism concentration in continuous cultures of filamentous microorganisms. *Journal of Applied Bacteriology* 30, 246-250.

ROELS, J.A., VAN DEN BERG, J. & VONCKEN, R.M. (1974). The rheology of mycelial broths. *Biotechnology and Bioengineering* 16, 181-208.

ROLINSON, G.N. (1952). Respiration of *Penicillum chrysogenum* in penicillin fermentations. *Journal of General Microbiology* 6, 336-343.

ROMANTSCHUK, H. & LEHTOMÄKI, M.(1978). Operational experiences of first full scale Pekilo SCP mill application. *Process Biochemistry* 13 (3), 16-17, 29.

ROWLEY, B.I. & BULL, A.T. (1973). Chemostat for the cultivation of moulds. *Laboratory Practice* 22, 268-289.

RUSHTON, J.H., COSTICH, E.W. & EVERITT, H.J. (1950). Power characteristics of mixing impeller. Parts I and II. *Chemical and Engineering Progress*. 46 (8), 395-404, (9), 467-476.

SMITH, J.M. & Van't RIET, K. (1976). Scale up for mass transfer in agitated gas-liquid reactors. Paper presented Third Annual Research Meeting, Institution of Chemical Engineering Salford, March 30-31st.

SMITH, J.M., Van't RIET, K. & MIDDLETON, J.C. (1977). Scale-up of agitated gas-liquid reactors for mass transfer. Second European Conference on Mixing, Cambridge, U.K. March 30th - April 1st. Paper F4, 51-66.

SOLOMONS, G.L. (1969). Materials and Methods in Fermentation. Academic Press Inc., London and New York.

SOLOMONS, G.L. (1975). Submerged culture production of mycelial biomass. In The Filamentous Fungi, Volume 1, pp 249-264 (Eds. Smith, J.E. and Berry, D.R.), Edward Arnold, London.

STEEL, R., LENTZ, C.P. & MARTIN,S.M. (1955). Submerged citric acid fermentation of sugar beet molasses: Increase in scale. *Canadian Journal of Microbiology* 1, 299-311.

STEEL, R. & MAXON, W.D. (1966). Studies with a multiple-rod mixing impeller. *Biotechnology and Bioengineering* 8, 109-116.

STEFANIAK, J.J., GAILEY, F.B., BROWN, C.S. & JOHNSON, M.J. (1946). Pilot plant equipment for submerged production of penicillin. *Industrial and Engineering Chemistry* 38, 666-670.

TAGUCHI, H. (1971). The nature of fermentation fluids. *Advances in Biochemical Engineering* 1, 1-30.

VAND, C. (1948). Viscosity of solutions and suspensions. Part I. Theory. *Journal of Physics and Colloid Chemistry* 52, 277-299.

VAN'T RIET, K. & SMITH, J.M. (1973). The behaviour of gas-liquid mixtures near Rushton turbine blades. *Chemical Engineering Science* 28, 1031-1037.

VAN'T RIET, K. & SMITH, J.M. (1975). The trailing vortex system produced by Rushton turbine agitators. *Chemical Engineering Science* 30, 1093-1105.

VAN'T RIET, K., BOOM, J.M. & SMITH, J.M. (1976). Power consumption impeller coalescence and recirculation in aerated vessels. *Transactions of Institution of Chemical Engineers* 54, 124-131.

CONSTRUCTION OF A LABORATORY-SCALE STIRRED TANK FERMENTER FOR FILAMENTOUS MICROORGANISMS

B. Kristiansen[1] and C.G. Sinclair[2]

[1]*Department of Applied Microbiology, University of Strathclyde, Glasgow G1 1XW*

[2]*Department of Chemical Engineering, UMIST, P.O. Box 88, Manchester M60 1QD*

INTRODUCTION

The construction and design of laboratory-scale stirred tank fermenters have been described in great detail (Ricica, 1966; Solomons, 1969; Evans *et al.*, 1970). They are produced commercially and the number of manufacturers is ever increasing. Traditionally the tower fermenter has represented the alternative to the stirred tank, but recently we have seen the introduction of new fermenter configurations. Among these are the thin channel fermenter (Gasner, 1974), the tubular loop fermenter (Ziegler *et al.*, 1977), the microbial film fermenter (Atkinson & Knights, 1975), the completely circular ring reactor (Lederach, Widmer & Einsels, 1978) and the rotating disc fermenter (Blain, Anderson, Todd & Divers, 1979). Current trends in fermenter design is to promote oxygen transfer and cell growth without the aid of mechanical agitation and to reduce the operating costs, as these fermenters illustrate. Because of its feasibility and reliability the stirred tank is still the preferred design (Kristiansen, 1978), particularly on a laboratory-scale where these criteria are more important than the operating cost.

Many of the stirred tanks available were designed to be used with yeasts or bacteria. In continuous culture the culture volume is kept constant using a constant overflow tube. With filamentous microorganisms, however, this is not a satisfactory method, acting as an internal cell feedback device (Solomons, 1972). The tube may also be blocked by growing mycelium (Bartlett & Gerhardt, 1959). Modification to the constant overflow system has met with some success (Righelato & Pirt, 1967). The overflow tube is employed as a combined air/broth outlet. A solenoid valve is used to close the outlet line at regular intervals, forcing the mycelial culture out on opening. A different approach to continuous cultivation of filamentous microorganisms, based on the load cell principle, is illustrated in this chapter. Control by weight measurements is used successfully on large scale fermenters and is finding application on laboratory fermenters. Using a load cell, the culture outlet point can be in the bottom plate rather than at the liquid surface.

It is difficult to purchase a fermenter which fits the process requirements exactly, but with the advance of modular design it has become possible to obtain fermenters carrying a bare minimum of modules. They are still costly pieces of equipment, however, and the alternative is to build. This is basically a matter of assembling items readily available although a few parts must be manufactured to order. By building, rather than buying, the fermenter can be tailored to any particular process and at the same time represent a considerable saving (Kristiansen, 1979). A number of design criteria, irrespective of the origin of the fermenter are illustrated below.

DESIGN CONSIDERATION

It is important to realise the constraints put on the fermenter design by process demands (Blakebrough, 1973). A continuous culture system must be designed to satisfy several important conditions. These have been outlined by Evans *et al.* (1970) and are given below:

(i) the culture is protected from contaminating microorganisms and must be enclosed,
(ii) the flow of fresh sterile medium is maintained at a constant, but variable rate,
(iii) the culture volume remains constant,
(iv) the culture is well mixed and satisfies the condition of C.S.T.R.,
(v) the dissolved oxygen level is maintained above the critical value by aeration and agitation in aerobic cultures,
(vi) the environment must be controlled.

A fundamental design objective is to make the equipment as flexible as possible, i.e. a modular design where each module performs one complete function and can easily be modified or replaced without affecting the rest of the structure.

The minimum size of a fermenter is largely defined by its ability to contain all the control equipment. Small fermenters will have a low medium demand in continuous culture, reducing the workload. Fungal batch processes are lengthy, however, requiring many samples. This should not affect the mixing characteristics of the broth. The initial and final culture volume should not differ by more than 10% (Solomons, 1971). An operating volume of less than 5 l may, therefore, restrict the versatility of the design.

The factors governing the choice of materials of construction have been outlined by King (1971):

(i) All materials coming into direct contact with the solutions entering the fermenter or the actual culture, must be corrosion resistant to prevent trace metal contamination.
(ii) The materials must be non-toxic, so that slight dissolution of materials or components does not inhibit culture growth.
(iii) The materials of the fermenter must withstand repeated sterilization with 15 psig steam.
(iv) The fermenter stirrer gland, ports and end plates must be easily machinable and sufficiently rigid not to be deformed or broken under mechanical stress.
(v) Visual inspection of the medium and culture is advantageous, transparent materials should be used where it is feasible.

CONSTRUCTION

The Growth Vessel

Borosilicate glass sections (Q.V.F.) with end plates are used extensively as growth vessels. Stainless steel vessels have found some application (Margaritis & Wilke, 1978), but the construction cost is prohibitive. The end plates can be stainless steel made to specification En 58J which will tolerate low pH values. The plates should carry as many inlet/outlet and process ports as practicable. Ports carrying process probes can be fitted with neoprene 'O' rings to allow the degree of immersion in the culture to be varied. A number of manufacturers can supply standard couplings and hose connections for the various ports. It is important, however, that all metal parts which are likely to come into contact with the culture medium are manufactured from the same grade stainless steel to prevent electrolytic corrosion.

Invariably, the interior of the fermenter will be covered with mycelium after a run (Bartlett & Gerhardt, 1959) and it will be necessary to dismantle the fermenter to clean it. The easiest way is probably to remove the top plate only. To do this, however, the bottom plate must support the whole weight of the growth vessel and the agitator motor must be mounted away from the top plate. This point is often overlooked by manufacturers, making it necessary to dismantle the whole vessel after a run. To form a seal between the plates and glass section, neoprene gaskets (Q.V.F.) and PTFE gasket sheaths (Q.V.F.) can be used. The sheath is strictly not necessary, but the gasket tends to adhere too strongly to the plate and glass section without it. Gaskets for the various ports can be made from neoprene or PTFE. Neither is ideally suited as they tend to rupture or deform under mechanical stress, but they are readily available and easy to cut.

Aeration and Agitation

In continuous culture, large variations in substrate, cell and dissolved oxygen concentrations will lead to control difficulties and prevent steady state being reached. It is necessary, therefore, that mixing is as perfect as possible and adequate rate of oxygen transfer is achieved in aerobic culture. The transfer must be potentially greater than the rate of oxygen utilization. To promote mixing it is common to employ one or more Rushton turbines and 4 baffles. The dimensions of the flat-bladed turbine and baffles used in fermenters (Aiba *et al.*, 1965) must be altered, however, to satisfy the criterion impeller diameter = ½ x vessel diameter, as specified for mycelial cultures (Solomons, 1969). In laboratory stirred tank fermenters it is common to use a liquid height to vessel diameter ratio of 1. One impeller will give adequate mixing when immersed to a depth of 2/3 of the liquid level. Clearly it may be necessary to alter its position, and most impellers are secured to the impeller shaft by a grub screw. The stirrer shaft is connected to the drive via a gland inserted through an end plate. Oil seals and teflon retainers will provide lubrication and allow steam sterilization. Alternatively, self-lubricating seal cartridges of graphite-impregnated teflon may be employed. A well designed stirred gland should not require cooling water. It has been suggested that the stirrer gland may offer a point of entry for contaminants and magnetically driven impellers have been introduced (Cameron & Godfrey, 1977). The impeller shaft is supported in ball-bearings and connected to the drive via magnets on either side of the end plate. This system is mechanically more complex, but gaining in popularity.

It is a temptation to install powerful agitator motors, hoping this will ensure perfect mixing. Using a 1/8 HP DC shunt wound motor (Norman Electrical Company Limited), we obtained good mixing in a 8.5 l filamentous broth with a dry weight of approximately 40 kg/m^3. The size of the motor is also dependant on the design of the stirrer gland as frictional forces may account for up to 75% of the power consumption in laboratory scale fermenters (Aiba, Humphrey & Millis,

1973). Normally a 1/4 h.p. motor should be sufficient for a 10 l fermenter.

To supply the air, a number of pumps are available, capable of delivering up to 25 l of air per minute. A wide range of flowmeters are also available. If the meter is fitted with a needle valve, it must be remembered that this is a control valve. To keep the air flow rate constant in spite of line pressure changes, Flostat control valves (G.A. Platon Limited) can be included in the air supply line. The flowmeter and control valve ought to be calibrated for approximately the same range. To filter the air miniature bacterial line filters (Microflow Limited) will remove submicron particles. The filters are steam sterilizable, but care must be taken to avoid wetting the filter membrane. A simple, but effective air filter can be obtained by filling a steam sterilizable container with non-absorbant cotton wool. Careful filling is necessary as channelling may reduce the efficiency of the filter.

If the air is supplied from a laboratory ring, it must pass through a primary filter and pressure reducer prior to the control valve. Mechanical filters can only remove water in a liquid form, and the amount of moisture in vapour form in compressed air is inversely proportional to the pressure. The primary filter should therefore be inserted at the part of the highest pressure, i.e. before the pressure reducer. Combined filter/regulators (C.A. Norgren Limited) are available. These are fitted with automatic water drain and can remove particles above 25 μm. They also carry pressure gauges which allows manual adjustment of the air supply pressure.

The design of the air sparger will influence the oxygen transfer. Generally, spargers producing small air bubbles promote high oxygen transfer rates. On a laboratory scale operating costs are normally not important, however, and high transfer rates can be achieved by rapid mixing. It may be sufficient, therefore, to let the air enter the broth via a tube sparger although a sintered disc or ring sparger will give higher oxygen transfer rates (Clark & Lentz, 1963).

Evaporation can be significant, especially in the prolonged batch runs often carried out with filamentous fungi. This is compensated for by using air saturated with water or a condenser at the air outlet. The former method may encourage contamination, the condenser will prevent the exit air filter being saturated with water. The filter is required to remove spores etc. A flow diagram for the air supply is included in Fig. 1.

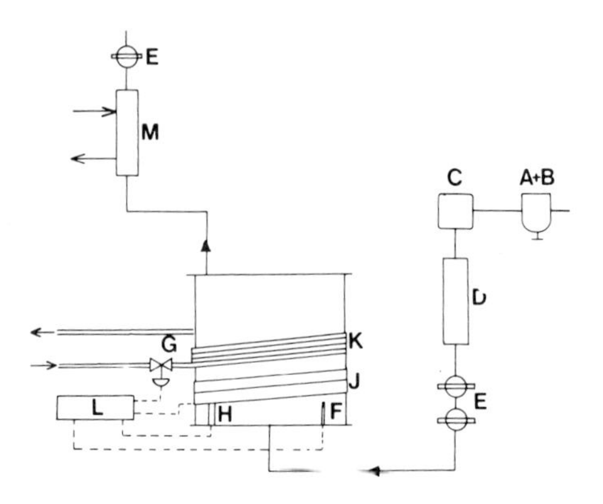

Fig. 1. Flow diagram for temperature control and air supply. Air: A + B - Combined filter/regulator (C.A. Norgren Limited); C - Control valve (G.A. Planton Limited); D - Flowmeter; E - Air filters (Microflow Limited); F - Temperature probe; G - Solenoid valve (Alexander Control Limited); H - Immersion heater (Bray Chromalux); J - Heating tape (Hotfoil Limited); K - Cooling coil (Orm Scientific Limited); L - Temperature controller (West Guardian); M - Condenser

Temperature Control

The capacity of the control elements depends on the cooling water, fermenter temperature and ambient conditions. A heating capacity of 300-400 W is normally sufficient for a 10 l fermenter. A cartridge type heating element encased in a stainless steel sheath inserted through the bottom plate is used extensively. At the end of a long batch run, mycelium will cover the heating element, forming a thick, charred crust. This can be removed easily, producing a slightly discoloured sheath. It can be avoided by reducing the fermentation time, but this is not necessary as the temperature control may not be adversely affected. To increase the heating capacity, a heating tape is wrapped around the fermenter vessel. Cooling is obtained by wrapping a cooling coil around the vessel, silicone rubber will suffice, but 'Calorex' D-shaped tubing (Orme Scientific Limited) is better as this is designed for heating and cooling purposes. Some fermenters employ hollow baffles or draught tube to circulate the cooling water. This ensures a large surface area for heat transfer. It makes the design more rigid, however, and the baffles or draught tube tend to be rather bulky. The fermenter space taken up by heating control elements must be kept down to maximise the volume available for the culture. It is also important to minimise the surface area available for microbial attachment, which can become quite severe with filamentous fungi. Cooling elements used on some fermenters must be inserted through the bottom plate. They provide another collecting point for the mycelium. It must be mentioned, however, that wrapping a heating tape and cooling coil around the fermenter is not the best solution aesthetically, but it is functional and cheap.

Controlling the temperature of circulating water is also used for temperature control. This necessitates the use of an external heat exchanger and is probably only feasible in fermenters with internal circulating coils. No heating element is required inside the fermenter, however.

The temperature control can be carried out by heating against constant cooling. The heating element(s) is working against the constant flow of cooling water. This will dampen temperature fluctuations but it is an expensive mode of operation. A solenoid valve at the cooling water inlet can be used to give an on-off cooling-heating control action. Fermenters have normally a slow response and a good controller will keep the fluctuations to ± 0.15°C or better. There are a number of excellent temperature controllers available, the majority support a temperature indicator. This is not essential but very helpful. The range of the indicators does not normally stretch to 121°C or above which is required for sterilization purposes. As most media are not sterilized *in situ* in laboratory fermenters this may not be too important. With the expanded temperature range the cost of the indicator/controller may be unnecessarily high. Provisions for manual override to give constant heating or cooling are also important, although they are not essential control specifications. A resistance bulb mounted in a stainless steel sheath inserted through the bottom plate is used extensively as a temperature sensing probe with very good results. The flow diagram in Fig. 1 includes a temperature control system.

Level Control

In a chemostat, steady state can be achieved by using a constant overflow tube to keep the culture volume constant. The problems encountered by using filamentous fungi have been illustrated above. Steady state can also be achieved, however, by keeping the culture mass constant. The load cells employed on large fermenters are not sufficiently sensitive to be used on laboratory fermenters, but the industry has realized that this is the best way to control cultures of filamentous microorganisms (Knopfel & Müller, 1978). A device for detecting culture mass in laboratory scale fermenters has been described (Brown & Patel, 1971) and later modifications have produced a reliable mass detector which can be sterilised *in situ* (Brown, unpublished; Kristiansen, 1976). The culture mass is detected by a differential pressure transducer connected between the top and bottom plates of the fermenter. The transducer has been described in detail (Kristiansen, 1976). Basically it is a U-tube manometer with one leg being the fermenter and the other being an external connection to the top plate. The external leg is separated from the fermenter contents by a steam sterilizable silicone rubber membrane thus allowing sterilization *in situ*. The increase in culture mass caused by feed flowing into the vessel will cause an increase in the fluid level in the external leg of the U-tube. This is detected by a level monitor (Fisons Limited) which triggers an air operated pinch valve in the outlet line, allowing broth to run out until the original culture mass has been retained. The pinch valve is essentially a piece of silicone rubber tubing, the flow through which is prevented by constricting the tube by compressed air (Brown & Inkson, 1972). The tubing is contained in a steel tee-piece which constitutes the valve body, as illustrated in Fig. 2. The air flow supply to the pinch valve is controlled by a three-way solenoid valve (Bellows-Valvair Limited). Triggered by the controller, the solenoid valve cuts the air supply and vent the pinch valve to the atmosphere.

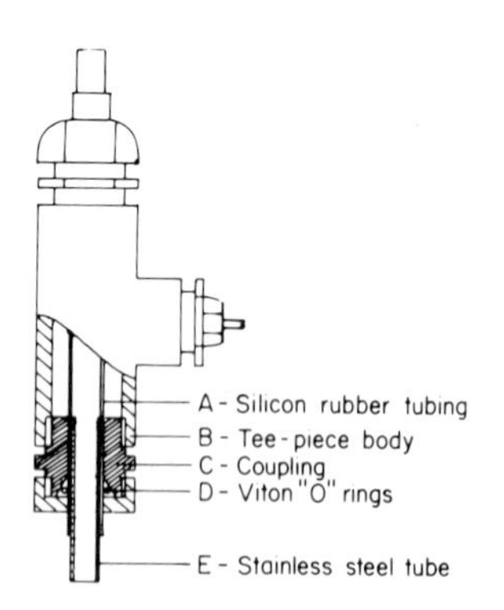

Fig. 2. Pinch valve (Brown & Inkson, 1972).

The outlet point is at the bottom plate and the flow through the valve is unrestricted. The diameter of the outline is 10 mm to avoid blocking by the mycelium. Samples of the outflow will be an accurate representation of the vessel contents. It has been demonstrated that this design can be used both in batch and continuous culture with filamentous fungi (Brown & Zainudeen, 1977; Kristiansen & Sinclair, 1978, 1979). It must be said, however, that the controller can be temperamental. Burst membranes and difficulties with the level monitor control setting are common teething problems. There is also considerable wear on the silicone rubber in the pinch valve and replacing it is tedious. A constant overflow tube is much simpler, but with filamentous microorganisms there tends to be no choice. A flow diagram is shown on Fig. 3.

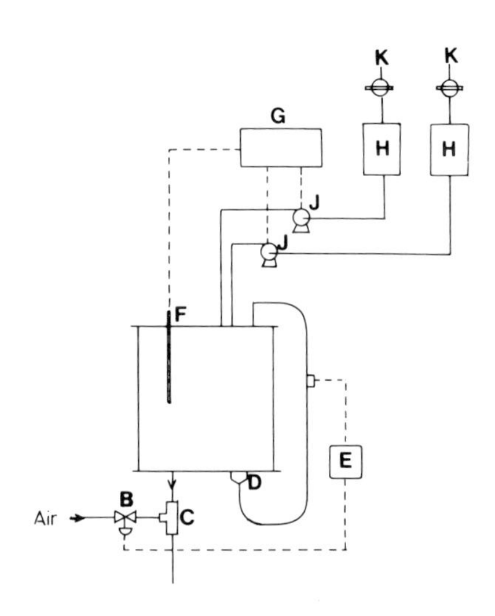

Fig. 3. Flow diagram for pH and level control. B - Three way solenoid valve (Bellows-Valvair Limited); C - Pinch valve; D - Pressure transducer; E - Level monitor (Fisons Limited); F - pH probe; G - pH indicator/controller (Analytical Measurements Limited); H - Titrant reservoirs; J - Peristaltic pump (Schuco Limited); K - Air filter.

pH Control

The pH of the environment affects growth and metabolite production of microorganisms and pH control is an essential part of research fermenters. A number of suitable pH combined indicator-controllers are manufactured. The basic difference lies in the price; a relatively inexpensive unit may be adequate. Provision for temperature compensation is important but digital display is not, although it looks good. The pH probes are invariably combined glass/reference electrodes. Steam-sterilizable probes are available in case sterilization *in situ* is essential. These probes tend to deteriorate with repeated sterilization, however, it may be sufficient to sterilize a probe by keeping it immersed in 70% alcohol (methanol or ethanol) prior to introducing it into the culture.medium.

The control system must incorporate addition of both acid and alkali although only one may be required theoretically. To supply the titrants, small peristaltic pumps (Schuco Limited) are available. The pumps are cheap, but excellent for this purpose. The flow rate can be adjusted by using different size silicone rubber tubing across the pumps, but this has its limitations because of the size of the nipples connecting the titrant lines to the pump tubing. Many pH controllers have energy regulators which can be used to vary the percentage on time of the pump from 0 to 100% for a continuous signal from the controller. This may also be used to control the titrant flow rate. If the flow rate makes pH control difficult, it may be easier to change the molarity of the titrant.

Leaks often occur where the supply lines are connected to the pump nipples. Using a supply line of silicone rubber or another suitable soft inert material and an 'O' to make a tight seal between the nipple and supply line will overcome this. If the pH indicator/controller is not supporting a recorder, it should have a recorder output. Quite often pH is an important process parameter and recording the pH can be a great help in evaluating the data. A flow diagram for pH control is included in Fig. 3.

Foam Control

Foaming may be a serious problem in fermentation broths. It is combatted by adding antifoam or using mechanical foam breakers. Antifoam may adversely affect the metabolic activity of the microorganisms or the broth consistency, causing product recovery problems. If the antifoam is metabolised, addition at regular intervals is required. It is important to avoid adding too much antifoam as this will reduce gas hold-up, lowering the oxygen transfer rate. A foam control system based on the addition of antifoam must, therefore, be very flexible in its mode of addition. We use a system of the following design. An adjustable height stainless steel conductivity probe is inserted through the top plate of the fermenter, detecting the foam. This triggers the foam controller to actuate a small peristaltic pump for a pre-set period of time, delivering a small quantity of antifoam. After a shot of antifoam has been delivered, there is a pre-set inhibition time during which the controller cannot be actuated. This is to prevent too much antifoam being added, as there is always a lag between foam detection and breakdown. The controller has also an override switch to deliver antifoam in the absence of a foam signal. This has two positions, 'one-shot' which delivers a single shot of antifoam when depressed and 'continuous' which will cycle the pump indefinitely at a rate dependent upon pump on-time and the inhibit time. This may seem excessive, but it is probably the only way of ensuring a minimum of antifoam is added to control the foam level.

Mechanical foambreakers can be used where addition of antifoam is not possible. These vary from an impeller situated above the liquid level to sophisticated gas/liquid separation devices. Some fermenters have foam breakers fitted with its own motor. These devices are excellent, but it is hard to justify the additional expense for a laboratory fermenter.

Sterilization

Provision must be made for sterilization *in situ*. The steam line with its valving arrangement must be designed to sterilize the vessel, feed line and air line simultaneously or independently. Thus, for example, the medium feed line can be steam sterilized at any time without affecting the culture. The titrant lines may not need sterilization, depending on the titrant. It is sufficient to run some titrant into the vessel prior to the sterilization cycle. The steam line or fermenter vessel must support a pressure indicator and a removable explosion shield must be used during sterilization. If process probes are sterilized *in situ*, they may require some time to settle. At the end of the sterilization cycle, air is supplied to the fermenter to ensure a positive pressure. Maintaining a positive pressure it is possible to open ports to insert probes during a run without contaminating the culture. This technique requires experience to be carried out successfully and has led to many abandoned runs.

CONCLUSIONS

Filamentous microorganisms pose additional demands to those associated with growth of unicellular microbes on fermenter design, particularly in continuous culture. The problems associated with the constant overflow tube can be solved by employing a design based on the load cell principle and an outlet point in the fermenter base plate. Constructing a fermenter can be rewarding. It is to be hoped that this chapter may prove useful to those who may wish to design and construct their own fermenter, while also giving a deeper knowledge for meaningful purchasing of a fermenter from one of the many available.

REFERENCES

AIBA, S., HUMPHREY, A. & MILLIS, N.F. (1973). *Biochemical Engineering*, London, U.K., Academic Press.

BARTLETT, M.C. & gerhardt, P. (1959). Continuous antibiotic fermentation. *Journal of Biochemical and Microbial Technology and Engineering* 1, 359-377.

BLAIN, J.A., ANDERSON, J.G., TODD, J.R. & DIVERS, M. (1979). Cultivation of filamentous fungi in the disc fermenter. *Biotechnology Letters* 1, 269-274.

BLAKEBROUGH, N. (1973). Fundamentals of fermenter design. *Pure and Applied Chemistry* 36, 305-315.

CAMERON, J. & GODFREY, E.I. (1979). Magnetic drives. *Biotechnology and Bioengineering Symposium* No. 4, 821-835.

BROWN, D.E. & INKSON, M.B. (1972). An air-operated pinch valve for chemostat outlet flow control. *Biotechnology and Bioengineering* 14, 1045-1046.

BROWN, D.E. & PATEL, A.R. (1971). Control of culture mass in a chemostat. *Biotechnology and Bioengineering* 13, 335-336.

BROWN, D.E. & ZAINUDEEN, M.A. (1977). Growth kinetics and cellulase biosynthesis in the continuous culture of *Trichoderma viride*. *Biotechnology and Bioengineering* 19, 941-958.

CLARK, D.S. & LENTZ, C.P. (1963). Submerged citric acid fermentation of beet molasses in tank type fermenters. *Biotechnology and Bioengineering* 5, 193-199.

EVANS, C.G.T., HERBERT, D. & TEMPEST, D.W. (1970). Continuous cultivation of microorganisms. 2. Construction of a chemostat. *Methods in Microbiology* 2, (Eds. J.L. Norris & D.W. Ribbons), pp 277-327. London, U.K., Academic Press.

KING, W.R. (1971). Static and dynamic modelling of systems for continuous cultivation of microorganisms. Ph.D. thesis, UMIST.

KNOPFEL, H.P. & MÜLLER, F. (1978). Methods for achieving steady-state conditions in laboratory fermenters. *Paper presented at First European Congress of Biotechnology*, Interlaken, Switzerland, 1978.

KRISTIANSEN, B. (1976). Production of organic acids by continuous culture of fungi. Ph.D. Thesis, UMIST.

KRISTIANSEN, B. (1978). Trends in fermenter design. *Industry and Chemistry*, 21 October 1978, pp 787-789.

KRISTIANSEN, B. (1979). Design of a fast flow reactor for production of SCP from flow wastes. *Paper presented at COST-Workshop on production and feeding of Single Cell Protein*, Jülich, West Germany, 1979.

KRISTIANSEN, B. & SINCLAIR, C.G. (1978). Production of citric acid in batch culture. *Biotechnology and Bioengineering* 20, 1711-1722.

KRISTIANSEN, B. & SINCLAIR, C.G. (1979). Production of citric acid in continuous culture. *Biotechnology and Bioengineering* 21, 297-315.

LAEDERACH, H., WIDENER, F. & EINSELE, A. (1978). Influence of the hydrodynamic on the productivity of bioreactors. *Paper presented at First European Congress of Biotechnology* Interlaken, Switzerland, 1978.

MARGARITIS, A. & WILKE, C.R. (1978). The rotofermenter. 1. Description of the apparatus, power requirements, and mass transfer characteristics. *Biotechnology and Bioengineering* 20, 709-726.

PUHAR, E., KARRER, D., EINSELE, A. & FLECHTER , A. (1978). Design and behaviour of completely filled bioreactors. *Paper presented at First European Congress of Biotechnology*, Interlaken, Switzerland, 1978.

RICICA, J. (1966). Technique of continuous laboratory cultivations. *Theoretical and Methodological Basis for Continuous Culture*. (Eds. I. Malek & Z. Fencl). pp 157-313. Prague, Czechoslovakia, Academic Press.

RIGHELATO, R.C. & PIRT, S.J. (1967). Improved control of organism concentration in continuous culture of filamentous microorganisms. *Journal of Applied Bacteriology* 30, 246-250.

SOLOMONS, G.L. (1969). *Materials and Methods in Fermentation*. London, U.K., Academic Press.

SOLOMONS, G.L. (1972). Improvements in the design and operation of the chemostat. *Journal of Applied Chemistry and Biotechnology* 22, 217-228.

RHEOLOGY OF MYCELIAL FERMENTATION BROTHS

G.W. Pace

Biospecialties, Penrhyn Road,
Knowsley Industrial Estate,
Prescot, Merseyside L34 9HY

INTRODUCTION

The performance of an industrial fermenter, determined by the maximum rates of heat and mass transfer achievable, is often controlled by the rheology of the fermentation broth. The rheological characteristics of culture fluids may also be important in determining the rates and methods of handling these fluids in various recovery unit operations, for example, liquid/cell separation. Rheological measurements can also be used to indirectly monitor morphological changes in mycelial cultures, which may be associated with changes in specific rates of product formation or other metabolic rates. In some fermentations where morphology remains constant, rheological measurements may provide a rapid method for determining cell concentration. In this paper, following a brief survey of basic rheology and the essential differences in the rheology of fermentation broths, viscosity measurements and the rheological characteristics of mycelial fermentations are discussed.

BASIC RHEOLOGICAL BEHAVIOUR

Rheology is the study of the deformation and flow behaviour of solids and fluids. Three basic types of ideal rheological behaviour (elastic, plastic and viscous) can be defined and visualised in terms of mechanical analogues (Fig. 1) (Charm, 1971; Muller, 1973; Sherman, 1970). Elastic behaviour is represented by a spring which stretches or deforms reversibly when a stress (force/unit area) is applied. The amount by which the spring stretches (α) is directly proportional to the applied stress (ξ) and the proportionality constant, E, is called the elasticity (Fig. 1). Plastic behaviour is modelled by a friction element to which a certain stress must be applied before the element will move and the stress at which movement takes place is known as the yield stress (ξ_0) (Fig. 1). The viscous behaviour of an ideal or Newtonian liquid is represented by a dash pot. On application of a force to the piston of the dash pot, the resultant rate of change of movement of the piston is directly proportional to the applied stress, and the proportionality constant, μ, is called the viscosity (Fig. 1).

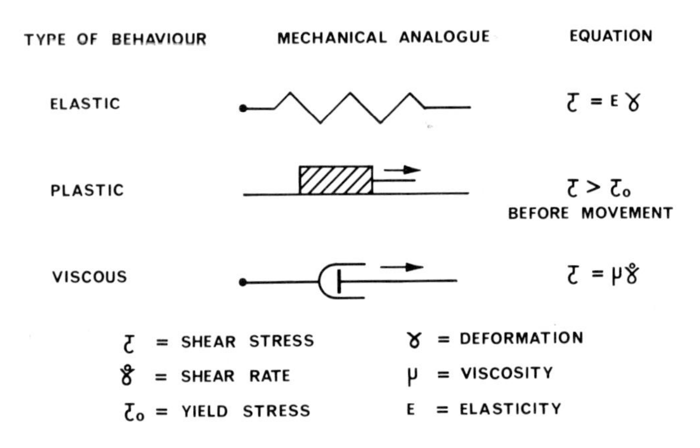

Fig. 1. Basic forms of ideal rheological behaviour.

The rheological behaviour of biological materials is usually non ideal and some examples of non ideal behaviour observed in biological materials are given below (Muller, 1973; Sherman, 1970; van den Temple, 1977).

i. *Non-Newtonian* flow is viscous flow in which the viscosity is not constant, and is a function of the rate of deformation of the fluid.

ii. *Viscoelastic behaviour* is a combination of elastic and viscous flow effects, on which other non ideal behaviour may also be imposed. One effect which is attributed to the elastic properties of viscoelastic liquids is normal stress. Normal stresses occur normal to the direction in which the applied shear stress is acting, and if present may lead to phenomena such as stirred polymer solutions climbing the shaft of the stirrer (Weissenberg effect) and reversal of flow patterns at certain Reynolds numbers in stirred systems.

iii. *Time dependent* behaviour occurs when the viscosity of

a fluid, sheared at a constant shear rate, changes with time.
iv. *Elongational or extensional* flow results when energy is dissipated by stretching of macromolecules, due to a velocity gradient existing in the direction in which the fluid is flowing.

Most literature on the rheological behaviour of mycelial fermentation broths appears limited to *in vitro* studies of steady shear viscous behaviour, whilst the significance of other rheological phenomena to transport mechanisms in the fermenter has received no attention. Steady shear viscous flow of a fluid can be appreciated by considering two parallel elements in the fluid separated by a distance dx (Fig. 2). If a force (F) is applied to the upper element of fluid, of area (A), causing it to move at a relative velocity (dv), then a gradient in velocity exists across the distance (dx). When the fluid is ideal or Newtonian, the shear stress of force per unit area (F/A) applied to this element is directly proportional to the velocity gradient or shear rate ($\dot{\gamma}$), and proportionally constant, (μ), is called the viscosity (Fig. 1 and 2). For a non ideal or non Newtonian liquid, the value of the proportionality constant, or viscosity, in the shear stress/shear rate equation varies with the magnitude of the relative velocity of the upper element, i.e. the velocity gradient or shear rate between the elements, e.g. Fig. 3 curve ii.

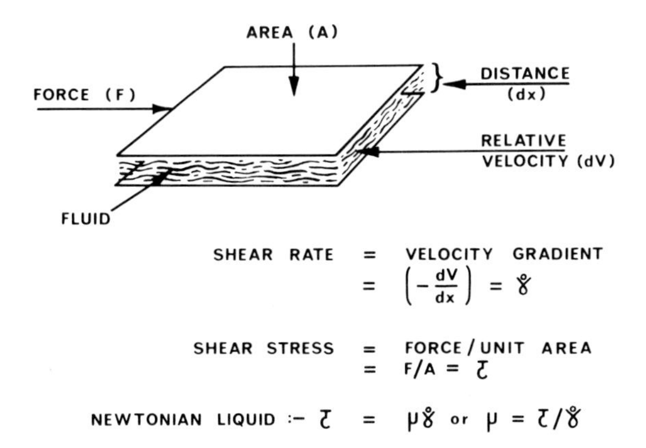

Fig. 2. Terms used in the definition of viscous flow behaviour.

One type of non Newtonian flow commonly observed is pseudoplastic viscous, or shear thinning flow, in which the viscosity or slope of the shear stresss-shear rate curve, decreases with increasing shear rate (Fig. 3, curve ii). Another form of non Newtonian behaviour is Casson viscous flow in which the fluid shows an apparent yield stress, and then behaves as a pseudoplastic liquid once the yield stress is exceeded (Fig. 3, curve iii). Thixotrophic behaviour, or time dependent pseudoplastic flow, occurs when the apparent viscosity of a pseudoplastic fluid decreases with time when it is sheared at a constant shear rate.

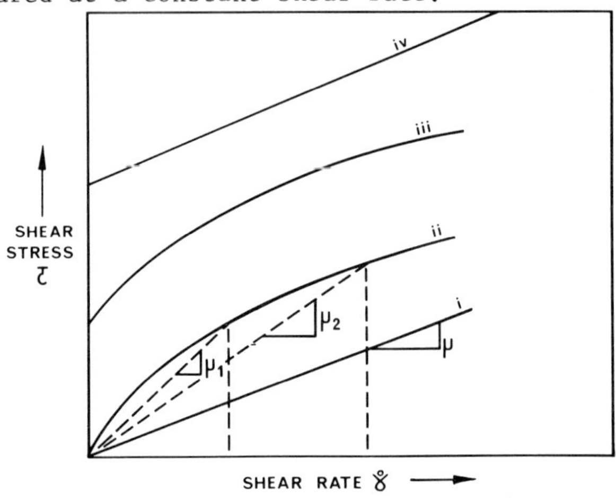

Fig. 3. Common types of viscous flow behaviour observed in biological fluids. i. Newtonian; ii. pseudoplastic; iii. Casson behaviour; iv. Bingham plastic.

RHEOLOGY OF FERMENTATION BROTHS

Fermentation broths can be grouped into three classes according to the phase (s) which determines their rheological properties (Table 1). In the first class, ideal or Newtonian behaviour is observed, and both the continuous (liquid) or discontinuous (cell) phases can contribute significantly to the rheology of the fluid. Examples in this class include bacterial and yeast culture fluids in which the cells behave as ideal spherical particles, having no preferred orientation in a shear field, and at high concentrations significantly increase viscosity (Deindoerfer & West, 1963; Perley, Swartz & Cooney, 1979; Shimmons, Svrcek & Zajic, 1976). In the next class, exemplified by mycelial fermentation broths, the rheology is typically complex and is caused by the presence of large, asymmetric, deformable particles, i.e. mycelial clumps, in an essentially Newtonian continuous phase. Tissue culture broths show similar behaviour (Kato, Kawazoe & Soh, 1978), and thus fall into this class. Rheological measurements on these broths appears to have been confined to steady-shear

Table 1. Rheological classification of fermentation broths.

Composition		Examples of Rheological Behaviour	Examples of Fermentation
Continuous Phase	Discontinuous Phase		
Aqueous plus* low molecular weight solutes	Single cells*	Newtonian	Bacterial, yeast
Aqueous plus low molecular weight solutes	Mycelium, pseudo-* mycelium, cell aggregates	non-Newtonian	Mould, streptomycetes, tissue culture
Aqueous plus* high molecular weight solutes	Single cells	non-Newtonian	Polysaccharide producing organisms

* Phase(s) determining rheology

measurements of viscosity, and indicate plastic, pseudoplastic and Casson behaviour (Fig. 3), although qualitative mention of other effects such as time-dependency can be found in the literature (Bongenaar, Kossen, Metz & Meijboom, 1973; Deindoerfer & West, 1960; Deindoerfer & Gaden, 1955; Karrow, Bartholomew & Sfat, 1953; Metz, 1976; Roels, van den Berg & Voncken, 1974; Sanchez-Marroquin, Ɖezma & Barreiro, 1971; Sato, 1961; Solomons & Perkin, 1958; Taguchi & Miyamota, 1966; Tuffile & Pinho, 1970). The third class of fermentations is that in which the continuous, or aqueous , phase is the principal determinant of the rheological behaviour of the broth. The type of rheological behaviour described in the literature for this class of fermentations includes non-Newtonian viscous flow and viscoelastic behaviour, including normal stress (Weissenberg) effects (Charles, 1978; Leduy, Marsan & Coupal, 1974; Miura, Fukushima, Sambuichi & Ueda, 1976; Pace, 1978).

Although it is often found that similar equations describe the non Newtonian viscous behaviour of fluids in the second and third classes given in Table 1, the mechanisms by which energy is dissipated in these types of fluids when they are stirred in a fermenter are different. These mechanisms, which can result in marked differences in heat and mass transfer rates are discussed further later.

VISCOSITY MEASUREMENTS OF MYCELIAL SUSPENSIONS

The particulate nature of mycelial suspensions poses special problems when attempting to measure their viscometric properties. In this section, the relative merits of the commonly used viscometers, such as the cone and plate, concentric cylinder, tube, and single spindle or infinite sea viscometers (Fig. 4) are discussed.

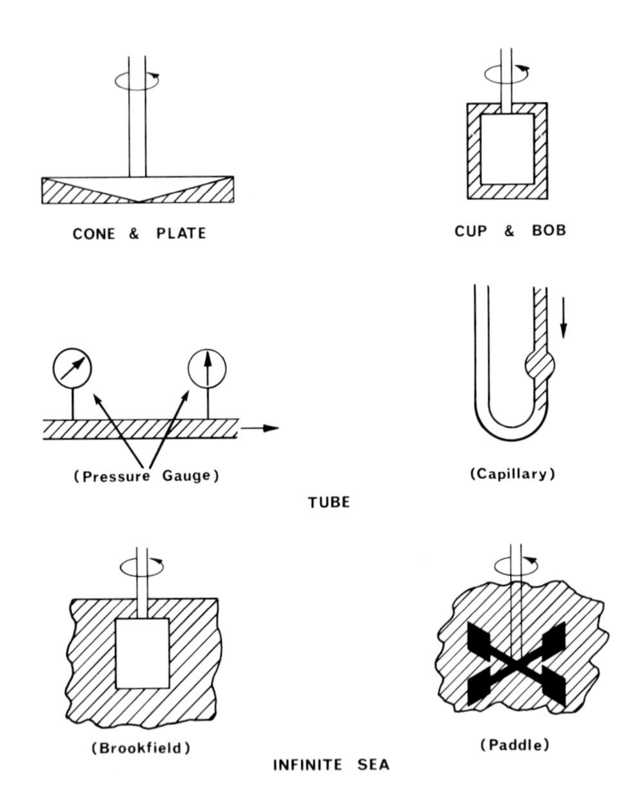

Fig. 4. Commonly used viscometers.

In the cone and plate viscometer the cone is usually rotated at a fixed speed relative to the plate, and the resultant torque on the cone or plate is measured. When the angle between the cone and plate is less than about $4^{o}C$, the shear rate in the fluid between the cone and plate is constant and independent of fluid properties, and is dependent only on rotational speed and the angle between the cone and plate. This viscometer, however, is normally unsuitable for measuring the viscosity of mycelial suspensions as the solids tend to jam between the cone and plate and lead to spuriously high readings.

In the concentric cylinder and cup and bob viscometers, the inner cylinder or bob is usually rotated at a fixed speed, and the resultant torque on the inner cylinder or bob is measured. When the gap between the two cylinders or the cup and bob is small, mycelial solids tends to compact and jam between the surfaces and this results in high readings (Cheng & Legrys 1975). Conversely, if the gap is increased to a point where solids no longer jam between the surfaces, other problems may arise. For example, solids may tend to settle in the viscometer (Bongenaar *et al.*, 1973), although this may be overcome by forced circulation normal to the direction of the shear field. Also, as the gap between the surface is increased the variation in shear rate across the gap increases and becomes a stronger function of fluid properties (van Wazer *et al.*, 1963). Slippage of solids at the instrument surfaces may also be a problem with these types of viscometers (Bongenaar *et al.*, 1973).

The tube viscometers are usually either one of two types: (i) the liquid flows through a tube of known diameter, under gravity and the time taken for a given volume to pass is measured (capillary viscometer); (ii) the liquid is pumped under laminar flow conditions through a tube of known length and diameter, and the pressure drop over the tube is measured.

Although the tube viscometers are simple and cheap they suffer from the same disadvantages of clogging, wall slip, and shear rate definition mentioned above. The gravity flow viscometers are usually designed for single point measurements, and therefore to define the non-Newtonian behaviour of liquids it is necessary to make measurements using a set of viscometers with varying diameters. Also, the use of these viscometers can be time consuming and tedious.

The infinite sea viscometers consist of a single surface rotating in a large volume of liquid. The common type of viscometer in this class is the single spindle Brookfield viscometer; however, problems of shear rate definition, wall slip, and settling of solids, make it poorly suited to measuring the rheology of mould suspensions. An adaption of this type of viscometer is the turbine viscometer in which a paddle is rotated at a fixed speed, in a large volume of liquid, and the resultant torque on the paddle is measured (Bongenaar *et al.*, 1973). This configuration is claimed to lessen the problem of wall slip, and solids settling, and is probably the best current approach to measuring the viscometric behaviour of mycelial suspensions. However, absolute measurements are difficult to obtain with this viscometer because of the problem of defining the shear rate at which the measurements are made, and false (high) readings, due to distortion of streamlines which leads to turbulent energy dissipation. An extention of this approach is the *in situ* measurement of viscosity made by determining the torque on the fermenter impeller when it is rotated at a sufficiently low speed to ensure laminar flow in the culture (Fewkes, 1977).

This approach may also be used to account for the presence of a gas phase on viscosity measurements (Fewkes, 1977).

Further information, giving the equations relating shear stress and shear rate to the dimensions of the various viscometer geometries, and fluid properties, and a fuller description of equipment is given in Garia-Gorras (1965), Sherman (1970) and van Wazer *et al.* (1963).

RHEOLOGY OF MYCELIAL FERMENTATION BROTHS

The rheological behaviour of mycelial suspensions arises from frictional and other energy losses which occur when suspensions are sheared. Mechanisms by which energy is dissipated include: purely viscous frictional losses within the liquid phase, and additional losses within the liquid phase due to distortion of streamlines by the particles; flexing of hyphae, resulting in distorted particle shapes; disruption of hyphae or mycelial flocs (Fig. 5).

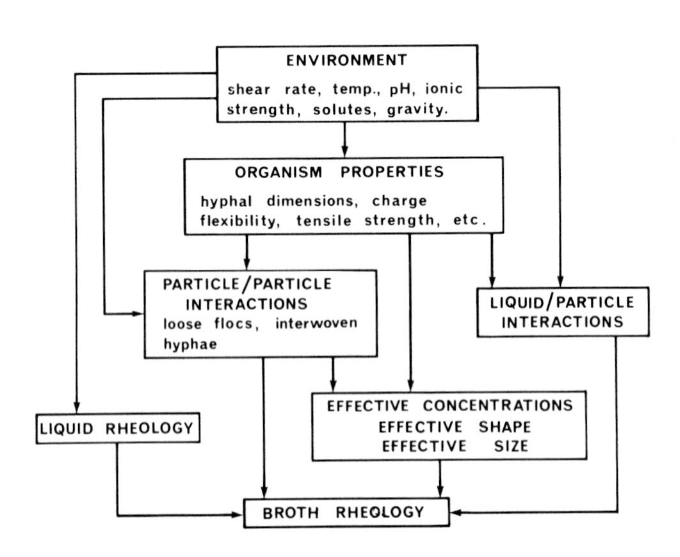

Fig. 5. Factors affecting the rheology of suspensions of mycelial microorganisms.

The effective size and shape, concentration, and size distribution of mycelial particles, primarily affects the amount of energy dissipated through the particle/liquid interactions and thus affects the measured rheology. Increasing the concentration of mycelium increases the volume fraction of particles present in the fluid, and thus the rheology of the suspensions. Some equations are given in the literature which describe the dependence of apparent viscosity on concentration or volume fraction (Caldwell & Babbitt, 1941; Blance & Bhavarju, 1976; Deindoerfer & Goden, 1955; Metz, 1976; Solomons & Weston, 1961), but because of the influence of other parameters, such as the effect of size and shape of particles on viscosity, they may not be generally applicable. One general feature, often noted, is that the relationship between apparent viscosity and concentration is non linear, and dilution of broth by say 10 to 15% may lead to a fall in apparent viscosity of about 50% (Chain, Gualandi & Marisi, 1966). This observation has obvious significance to the mode of operation of a fermenter when trying to maximise heat and mass transfer.

The effective or hydrodynamic size and shape of mycelial particles is determined by the size and shape of the mycelium, and the amount of liquid included and bound by the particle. The actual size and shape is dependent on a number of variables and interactions and includes:- (i) the properties of the mould hyphae, such as their dimensions, geometry, flexibility, tensile strength and charge; (ii) environmental variables which may affect growth conditions, and thus mould morphology, or those which may distort, disrupt or affect the flocculation of particles.

For example, at equal concentrations, filamentous suspensions are viscous and pseudoplastic and often show a yield stress by extrapolation), whereas suspensions of pellets are of lower viscosity and tend to approach Newtonian behaviour (Table 2) (Fig. 6) (Carilli, Chain, Gualandi & Morisi, 1961; Chain *et al.*, 1966). Suspensions containing short mycelia, compared with those containing long mycelia, typically show lower apparent viscosities (Chain *et al.*, 1966). Similar effects have been noted in plant tissue cultures of tobacco cells, in which a 27.5 fold increase in the apparent viscosity of the culture was correlated with the increase in cell mass and cell aggregate size. Metz (1976), in a recent study, found quantitative relationships between the size and shape of filamentous suspensions of two strains of *Penicillium chrysogenum* grown in batch culture, and steady shear viscosity measurements of these suspensions:

$$\tau c = 1.67 \times 10^{-4}\ Cm^{2.5}\ (Le^{+})^{0.8}$$

Table 2. Viscometric behaviour of mycelial suspensions

Type of behaviour	Equation
Pseudoplastic	$\tau = K (\dot{\gamma})^h$ where $n < 1$
Bingham or plastic	$\tau = \tau o + K_B (\dot{\gamma})$
Mixed or modified power law	$\tau = \tau o + K (\dot{\gamma})$ where $n < 1$
Casson	$\sqrt{\tau} = \sqrt{\tau_c} + Kc \sqrt{\dot{\gamma}}$

Notes: τ = shear stress; $\dot{\gamma}$ - shear rate; τo - yield stress; K = consistency index; n = flow behaviour index; K_B = Singham constant; τc = Casson yield stress; K_C = Casson constant; τo, τc also see Fig. 6.

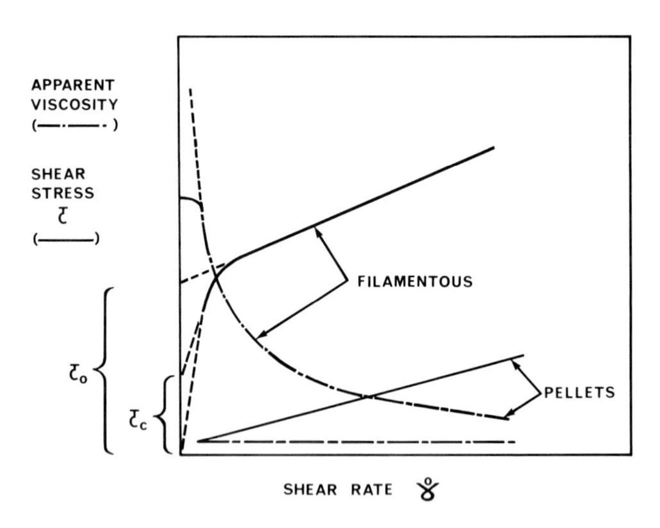

Fig. 6. Comparative viscous behaviour of pelleted and filamentous mycelial suspensions.

$$K_C = 1.37 \times 10^{-3} Cm^{1.0} (Lhgu)^{0.6}$$

τc = Casson yield stress ($N m^{-2}$); K_C = Casson constant ($N^{0.5} s^{0.5} m^{-1}$); C_m = concentration of mould ($kg m^{-3}$); Le^+ = length of main hyphae diameter of hyphae; Lhgu = total length of hyphae in particle/number of growth tips (m).

However, these relationships were not found to apply to the same strains grown under different conditions in continuous culture and Metz hypothesised that this was due to changes in hyphal flexibility with growth conditions. To demonstrate the influence of hyphal flexibility on rheological behaviour, Metz showed that the addition of sodium chloride to mycelial suspensions affected the Casson yield stress of suspensions, and claimed this was due to changes in cell turgor pressure via osmotic affects and thus flexibility. The shear rate or applied stress can directly change particle shape or size through distortion or disruption, or it can indirectly affect the mould morphology by influencing the growth conditions, such as dissolved oxygen availability, or carbon dioxide release. The disruption of mycelial particles is thought to be dependent on the applied stress rather than the velocity gradient or shear rate present in the liquid phase, and is enhanced by the fluctuations and increased stresses which occur in turbulent flow (Bhavaraju & Blanch, 1976; Taguchi, 1971).

Although the effect of size distribution on the rheology of mycelial broth does not appear to have been investigated, studies on other dispersed systems have shown it to affect viscosity. For example, when a suspension contains two different sized particles and the solid volume concentration is held constant, the apparent viscosity of the mixture passes through a minimum as the composition is varied (Sherman, 1970).

As discussed above, energy may be dissipated in the distortion or disruption of particles, but energy may also be dissipated in the disruption of flocs or aggregated particles. The presence of large and small aggregates has been observed in the fermenter with the average particle size increasing with distance from the impeller (Metz, 1976; Fewkes, 1977). It is likely that a different shear dependent equilibrium, between flocculation and deflocculation, exists throughout the fermenter. The position of the equilibrium will also depend on the properties of the microorganism, e.g. its charge and geometry, and the environment, e.g. pH and ionic strength. In a high shear environment the rate of destruction of flocs will be much higher than their rate of formation, and flocs will tend to the size of their unit components. At low shear rates larger flocs will occur and the influence of flocculation on the rheological properties of the suspension will increase. For example, increased flocculation may tend to enhance any thixotropic or time dependent pseudoplastic behaviour, and in the studies of Bongenaar *et al.* (1973) and Metz (1976) they noted the time dependency of steady shear viscosity measurements of mycelial suspensions which flocculated rapidly. Increased flocculation may promote the formation of solids structure at rest, or very low shear rates, and result in plastic flow behaviour. Several authors (e.g. Deindoerfer & West, 1963) have claimed that mycelial suspensions show plastic behaviour based on extrapolation of viscosity data measured at relatively high shear rates (Fig. 6). However, such evidence is insufficient to prove the existence of a true yield stress, i.e. a solids structure, at zero or low shear stresses.

The principal factors and interactions which may contribute to the *in situ* rheology of a mycelial fermentation broth are summarised in Fig. 7. The relative importance of these factors, such as the environment, medium, products and micro-organism, will vary with the type of fermentation and throughout the fermentation. For example, in the early stages of baker's yeast fermentation, a medium component, molasses, principally determines the broth rheology, but as sugar is utilised and cell mass is produced the cell content of the broth becomes the main factor. In the pullulan fermentation the filamentous yeast produces an extracellular polysaccharide, pullulan, which gives high broth viscosities. When the continuous (liquid) phase is the principal contributor to culture rheology the temperature will strongly influence the rheology. The influence of dispersed air on the viscometric properties of mycelial broths has received little attention in the literature. It could be hypothesised that the measured viscosity of a mycelial broth containing dispersed air could be either decreased by the presence of air, due to a dilution effect, or increased, if the air bubbles behave as rigid particles like other solids. In a recent study Fewkes (1977) compared *in situ* measurements of the viscosity of gassed *Streptomyces niveus* broths with *in vitro* measurements of degassed broths and found the *in situ* apparent viscosities to be about 1.5 times higher.

He claimed that the presence of this extra dispersed phase (*i.e.* air) in the *in situ* case contributed to the difference. However, differences in the geometry between the *in situ* "impeller" viscometer and the *in vitro* single cylinder viscometer may also partially account for the result. The observations of Bongenaar *et al.* (1973) also suggest the importance of air on mycelial broth rheology as they claim it is important to degass samples to obtain reproducible readings. In contrast to Fewkes, Charles (1978) claimed that in xanthan fermentations, air bubbles have only a small effect on broth viscosity.

Although simple viscosity measurements on mycelial broths indicate the amount of energy necessary to overcome the total friction in the system, the measurements in themselves may

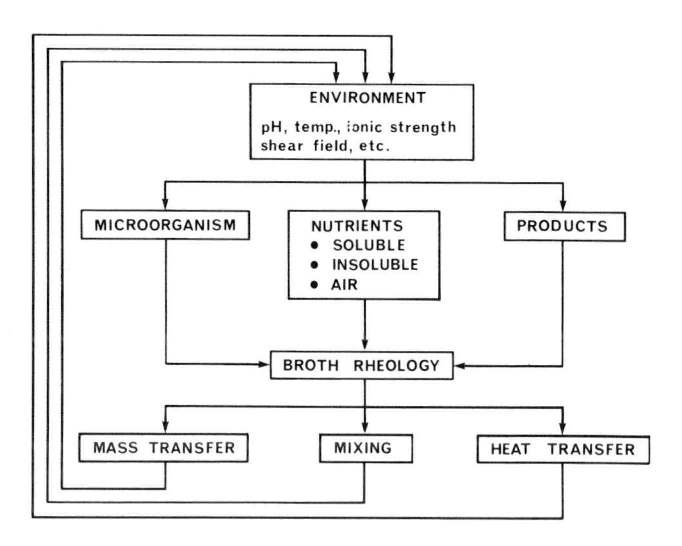

Fig. 7. Principal factors and interactions determining the *in situ* rheology of fermentation broths.

not be sufficient to establish the effect of rheology on mixing patterns, homogeneity and heat and mass transfer. This is well illustrated by some recent work in which heat transfer coefficient and steady shear viscosity measurements made in the laminar flow regime, on a mycelial *(Aspergillus niger)* broth and a microbial polysaccharide containing *(Xanthomanas campestris)* broth, were compared (Blakebrough, McManamey & Tart, 1978). The investigators found that although the heat transfer from the xanthan culture was modelled by a theoretical expression for heat transfer containing a modification for non-Newtonian flow, the mould suspension showed heat transfer coefficients up to 4 times those predicted by the equation. This was despite the fact that both broths were well modelled by an equation describing pseudoplastic viscous flow, and their results clearly show that some parameter other than viscosity is important in determining the heat transfer characteristics of the mycelial containing broth. Blakebrough *et al.* (1978) hypothesised that the enhancement of the heat transfer coefficient was due to the influence of the mycelial particles on the effective thickness of the boundary layer on the heat transfer surface. This work also demonstrates the invalidity of using solutions of polymers to simulate mycelial broth rheology in transport phenomenon studies.

CONCLUSIONS

Although measurements of the viscosity of mould suspensions are essential in describing quantitatively transport mechanisms which occur in fermenters, they are probably insufficient by themselves to fully describe the effect of rheology on the heat and mass transfer and mixing capabilities of a fermenter. Fruitful areas of research appear to be: (i) the determination of all the rheological or physical characteristics of the broth which are important in controlling transport phenomena, including the most meaningful way of determining viscosity; (ii) how the microorganism and its interactions combine to determine the broth rheology, and (iii) how to use our knowledge of rheological properties, such as shear thinning, to advantage in the design of a novel fermenter.

REFERENCES

BLAKEBROUGH, N., McMANAMEY, W.J. & TART, K.R. (1978). Heat transfer to fermentation systems in an air-lift fermenter. *Transactions Institution of Chemical Engineers* 56, 127-135.

BLANCH, H.W. & BHAVARAJU, S.M. (1976). Non-Newtonian fermentation broths: rheology and mass transfer. *Biotechnology and Bioengineering* 18, 745-790.

BHAVARAJU, S.M. & BLANCH, H.W. (1976). A model for pellet break-up in fungal fermentations. *Journal of Fermentation Technology* 54, 466-468.

BONGENAAR, J.J.T.M., KOSSEN, N.W.F., METZ, B. & MEIJBOOM, F.W. (1973). *Biotechnology and Bioengineering* 15, 201-206.

CARILLI, A., CHAIN, E.B., GUALANDI, G. & MORISI, G. (1961). Aeration studies. III. Continuous measurement of dissolved oxygen during fermentation in large fermenters. *Science Reports. 1st Super. Sanita* 1, 177-189.

CALDWALL, D.H. & BABBITT, H.E. (1941). Flow of muds, sludges, and suspensions in circular pipe. *Industrial and Engineering Chemistry* 33, 249-256.

CHAIN, E.B., GUALANDI, G. & MORISI, G. (1966). Aeration studies. IV. Aeration conditions in 3000 litre submerged fermentation with various microorganisms. *Biotechnology and Bioengineering* 8, 595-619.

CHARLES, M. (1978). Technical aspects of the rheological properties of microbial cultures. *Advances in Biochemical Engineering* 8 (Eds. T.K. Ghose, A. Fiechter & N. Blakebrough) pp. 1-59. Berlin West Germany: Springer-Verlag.

CHARM, S.E. (1971). *Fundamentals of Food Engineering*. Westport U.S.A.: AVI.

CHENG, D.C.H. & LEGRYS, G.A. (1975). The use of a coaxial cylinder viscometer on fermentation broths. *Annual Research Meeting of Institute of Chemical Engineers*. Collected papers - biochemical engineering.

DARBY, R. (1976). *Viscoelastic Fluids*. New York U.S.A.: Marcel Dekker.

DEINDOERFER, F.H. & GADEN, E.L. (1955). Effects of liquid physical properties on oxygen transfer in penicillin fermentation. *Applied Microbiology* 3, 253-257.

DEINDOERFER, F.H. & WEST, J.M. (1960). Rheological examination of some fermentation broths. *Journal of Biochemical and Microbial Technology and Engineering* 2, 165-175.

DEINDOERFER, F.H. & WEST, J.M. (1963). Rheological properties of fermentation broths. *Advances in Applied Microbiology* 2, 265-273.

FEWKES, R.C.J. (1977). Physical parameters governing the biological activity of a non-Newtonian mycelial fermentation broth. Ph.D. Thesis, M.I.T.

GARCIA-BORRAS, T. (1965). Calibrate rotational viscometers for non-Newtonian fluids. *Chemical Engineering* 72, 176-177.

KARROW, E.O., BARTHOLOMEW, W.H. & SFAT, M.R. (1953). Oxygen transfer and agitation in submerged fermentations. *Agricultural Food Chemistry* 1, 302-306.

KATO, A., KAWAZOE, S. & SOH, Y. (1978). Viscosity of the broth of tobacco cells in suspension culture. *Journal of Fermentation Technology* 56, 224-228.

LEDUY, A., MARSAN, A.A. & COUPAL, B. (1974). A study on the rheological properties of a non-Newtonian fermentation broth. *Biotechnology and Bioengineering* 16, 61-76.

METZ, B. (1976). From pulp to pellet. Ph.D. Thesis, Technological University, Delft.

MIURA, Y., FUKUSHIMA, S., SAMBUICHI, M. & UEDA, S. (1976). Relation between rheological properties of the broth and time course in the pullulan fermentation. *Journal of Fermentation Technology* 54, 166-170.

MULLER, H.G. (1973). *An Introduction to Food Rheology*. London U.K.: Heinemann.

PACE, G.W. (1978). Mixing of highly viscous fermentation broths. *The Chemical Engineer* November No. 338, 833-837.

PERLEY, C.R., SWARTZ, J.R. & COONEY, C.L. (1979). Measurement of cell mass concentration with a continuous flow viscometer. *Biotechnology and Bioengineering* 21, 519-523.

ROELS, J.A., van den BERG, J. & VONCKEN, R.M. (1974). The rheology of mycelial broths. *Biotechnology and Bioengineering* 16, 181-208.

SANCHEZ-MARROQUIN, A., LEDEZMA & BARREIRO, J. (1971). Oxygen transfer and scale-up in lysine production by *Ustilago maydis* mutant. *Biotechnology and Bioengineering* 13, 419-429.

SATO, K. (1961). Rheological studies on some fermentation broths. (IV). Effect of dilution rate on rheological properties of fermentation broth. *Journal of Fermentation Technology* 39, 517-520.

SHERMAN, P. (1970). *Industrial Rheology*. London U.K.: Academic Press.

SHIMMONS, B.W., SVRCEK, W.Y. & ZAJIC, J.E. (1976). Cell concentration control by viscosity. *Biotechnology and Bioengineering* 18, 1793-1805.

SOLOMONS, G.L. & PERKIN, M.P. (1958). The measurement and mechanism of oxygen transfer in submerged culture. *Journal of Applied Chemistry* 8, 251-259.

SOLOMONS, G.L. & WESTON, G.O. (1961). The production of oxygen transfer rates in the presence of mould mycelium. *Journal of Biochemical and Microbial Engineering Technology* 3, 1-6.

TAGUCHI, H. (1971). The nature of fermentation fluids. *Advances in Biochemical Engineering* 1 (Eds. T.K. Glose and A. Fiechter) pp 1-30. Berlin West Germany: Springer Verlag.

TAGUCHI, H. & MIYAMOTO, S. (1966). Power requirement in non-Newtonian fermentation broth. *Biotechnology and Bioengineering* 8, 43-54.

TUFFILE, C.M. & PINHO, F. (1970). Determination of oxygen transfer coefficients in viscous streptomycete fermentations. *Biotechnology and Bioengineering* 12, 849-871.

van den TEMPEL, M. (1977). The effects of elongational flow during processing on product rheology. *The Chemical Engineer* February 95-97.

van WAZER, J.R., Lyons, J.W., KIM, K.Y. & COLWELL, R.E. (1963). *Viscosity and Flow Measurement*. New York U.S.A.: Interscience.

INTERACTIONS BETWEEN FERMENTER AND MICROORGANISM: TOWER FERMENTER

R. Cocker

Glaxo Operations (U.K.) Limited,
Ulverston, Cumbria

INTRODUCTION

There are many examples in the literature of attempts to deal with the engineering aspects of the effect of fermenters on the mixing of microbial broths. It is now well established that the suspension viscosity developed during mould fermentations in stirred tanks often has a profound effect on mixing and mass-transfer. As an example, it is a key variable in published control algorithms (Ryu & Humphrey, 1973). However it is not often that we find descriptions of the ways in which fungal organisms can, by adaptation, modify the mixing and mass-transfer characteristics of fermentations. This paper outlines the ways in which it has been demonstrated that fungal organisms can affect the behaviour of fermentations in a particular type of fermenter *viz.* the tower fermenter.

INTERACTIONS BETWEEN MORPHOLOGY, RHEOLOGY AND FERMENTER

Batch culture

The first example illustrating the above interactions is a rheological study carried out on a 3.5 m high, 1 m^3 capacity tower fermenter during the batch fermentation of *Aspergillus niger*. The two fermentations concerned developed different morphologies; these will be referred to as Type I (Fig. 1) and Type III (Fig. 2).

The following graphs illustrate the interactions between morphology, rheology and fermenter performance. Note in Fig. 3, the gross difference between the absolute values of apparent viscosity for the two types, together with the

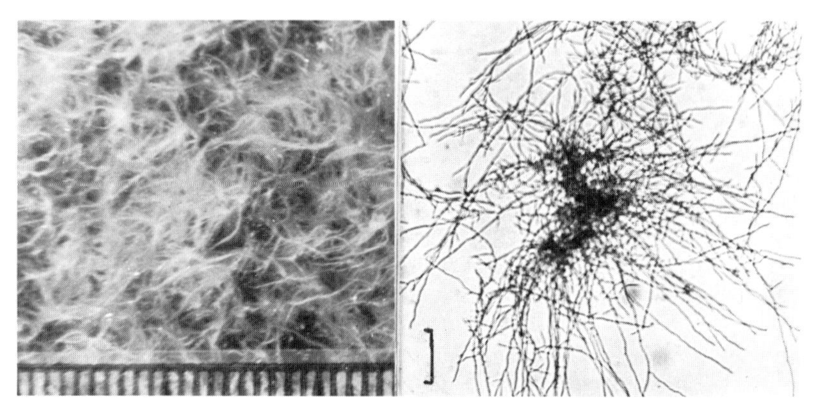

Fig. 1. Type I colonies (L.H. actual appearance; 1 division = 1mm. R.H. bright field - bar = 100 μm).

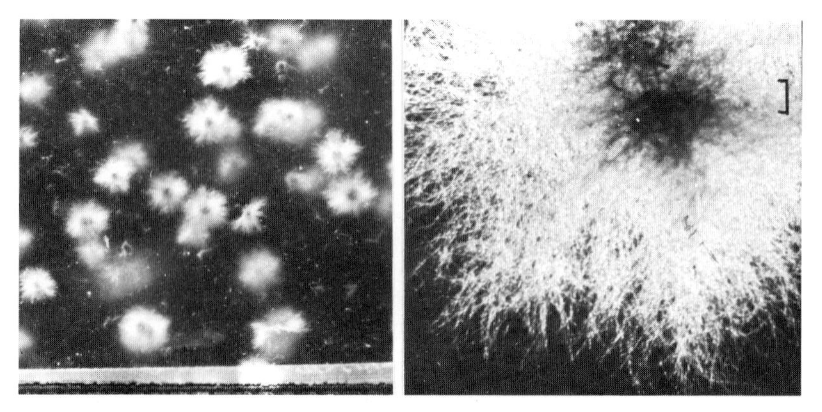

Fig. 2. Type III colonies (L.H. actual appearance; 1 division = 1 mm. R.H. dark field - bar = 100 μm).

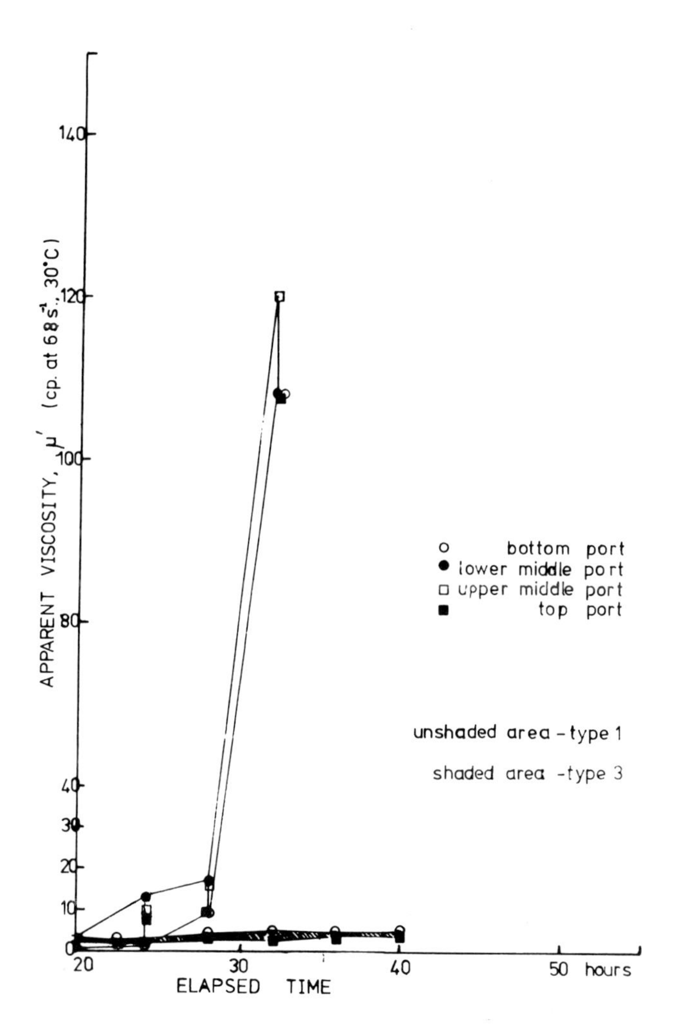

Fig. 3. Comparison of axial variation in apparent viscosity with time for type 1 and type 3 colonies.

differences in axial patterns of viscosity. Fig. 4 illustrates particularly how Type III colony morphology results in uniformly low values of viscosity with respect to both time and column height. Comparison of the dry weight profiles however (Fig. 5, 6) indicates that the biomass has differing axial distributions with respect to time and morphological type. For example, Type I (fig. 5) has a relatively even distribution of biomass within the columns. There is a slight

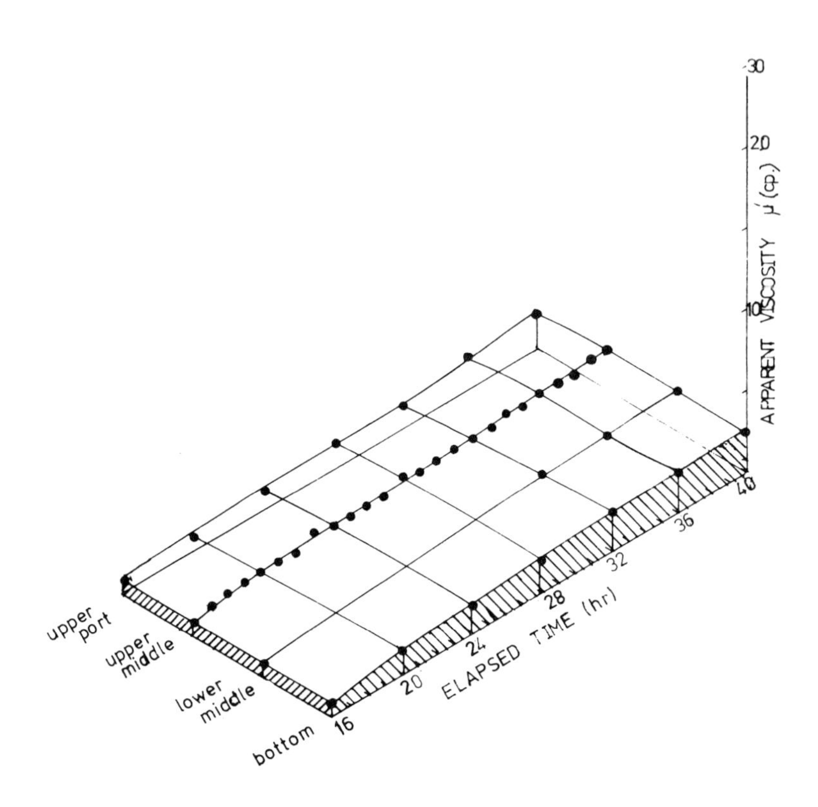

Fig. 4. Axial distribution-profile of apparent viscosity during a 450 l fermentation with Type III colonies.

tendency to have higher concentrations nearer the top of the reactor. Note that the absolute values of dry weight are lower than for Type III (Fig. 6), which has higher biomass concentrations in the middle zones of the column. The most dramatic difference however is the sudden reduction in growth seen with Type I after above 30h of fermentation (Fig. 5), at which point the broth behaves in a very turbulent manner with gas slugging. This turbulence appears to be associated with the rheological behaviour of the broth (Fig. 7). The apparent viscosity increases in a similar fashion for both types only up to about 22h at which point in the case of Type I the broth becomes non-Newtonian, at slightly elevated values of apparent viscosity. This is shortly followed by an abrupt transition to a second non-Newtonian phase in which average apparent

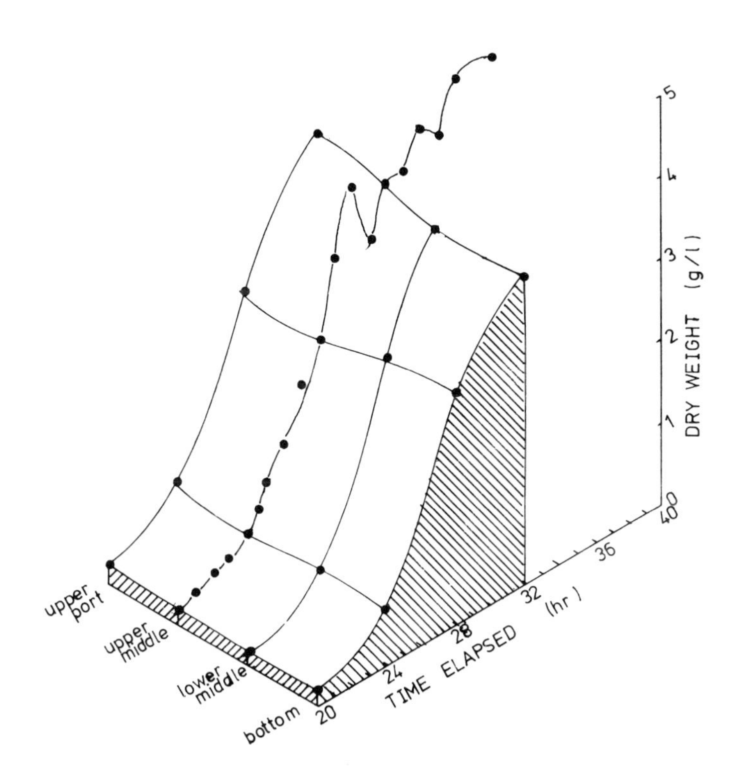

Fig. 5. Dry weight profile during a 450 l fermentation showing axial distribution of mycelium (Type I) in column.

viscosities are about four times greater than before. It is in this second phase that the mixing properties in the column deteriorate and the growth rate suffers.

Theoretical background

The fundamental mechanisms by which mould colonies affect the viscosity of their suspensions have not been considered in detail by many workers but some work has been carried out for simpler particulate suspensions. The first theoretical consideration of the rheology of particulate suspensions was that by Einstein (1911). He noted that the presence of small, rigid spheres in a liquid deformed flow-patterns. The suspension therefore has a higher viscosity than the supernatant liquid. He therefore determined that:

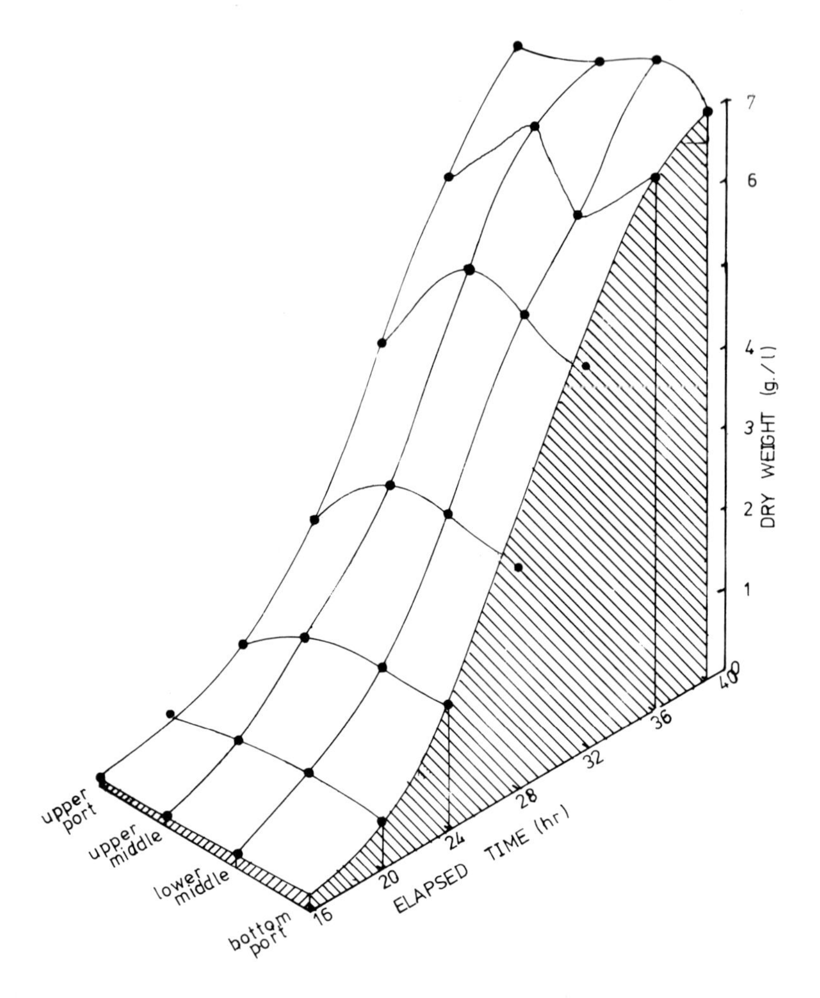

Fig. 6. Dry weight profile during a 450 l fermentation showing axial distribution of mycelium (Type III) in column.

$$\mu_r = \mu_s/\mu_o = 1 + 2.5\,\phi$$

where μ_r = viscosity ratio

μ_s = suspension viscosity

μ_o = liquid viscosity

ϕ = volume-fraction of solids plus any immobilised liquid.

This relationship applied only to dilute suspensions (ϕ is

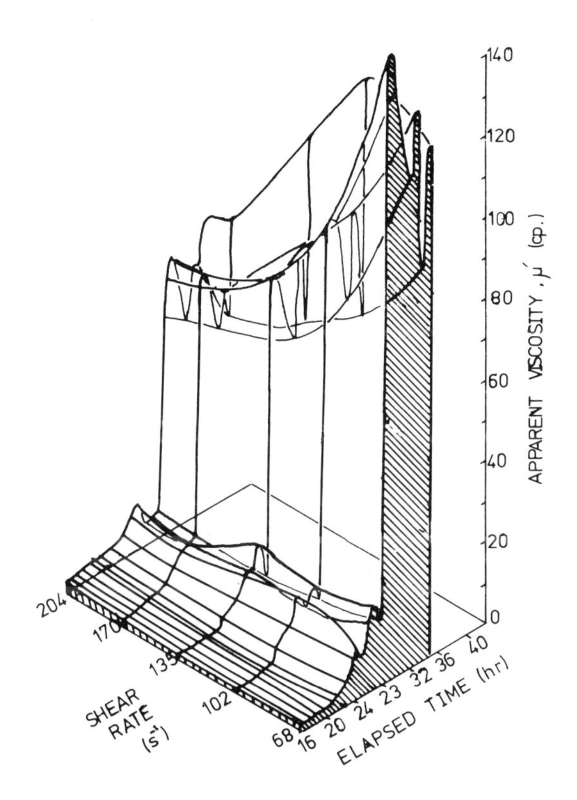

Fig. 7. Three-dimensional relationship between apparent viscosity, shear-rate , and time, for a 450 l fermentation involving Type I growth (isometric projection).

<0.05). This has been modified to account for higher particle concentrations. For $\phi > 0.08$, the following relationship has found to be valid.

$$\mu_r = \frac{(1 + 0.5\,\phi - 0.5\,\phi^2)}{(1 - 2\,\phi - 9.6\,\phi^2)}$$

Microbial particles, cells, or conglomerates often bear little resemblance to rigid spheres, *e.g.* Aiba *et al.* (1962) found that for dilute yeast suspensions the viscosity was less than that predicted by the formulae and they advanced the theory that the flexibility of the biological material reduced viscous drag such that:

$$\mu_r = 1 + \phi \quad (\phi < 2.5)$$

Measurements by James (1973) on *Aspergillus niger* M1 colonies of Type III from tower fermenters gave results which differed widely from Einstein's results for smooth spheres, the viscosity being considerably higher (Fig. 8). This suggests that the open structure of the spheres in some way increases viscous drag between colonies and between liquid and colonies.

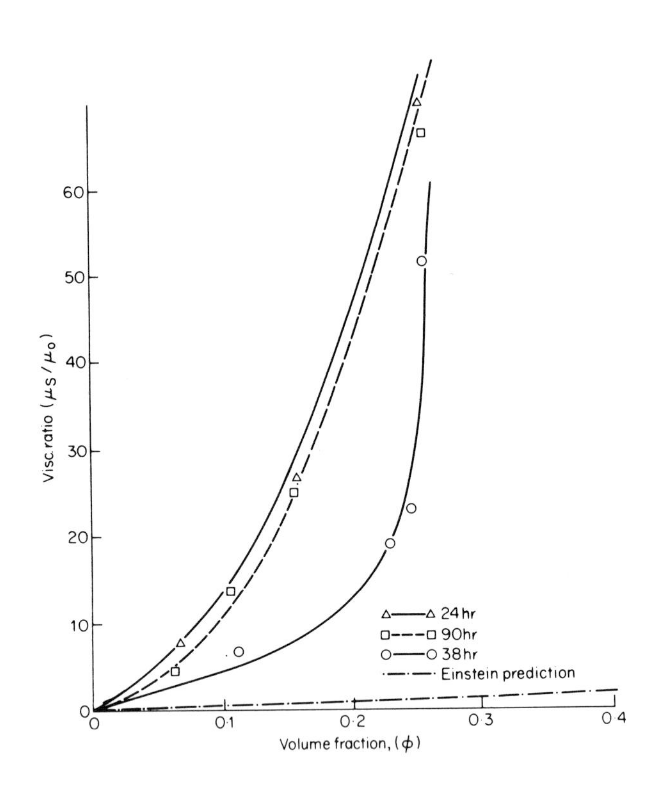

Fig. 8. Relationship of suspension viscosity to predicted viscosity for *A. niger* M1 (data of James, 1973).

The difficulty of developing theoretical models for such large particles is illustrates in this case by the sharp four-fold increase in apparent viscosity and shear-stress (Fig. 7), followed by a random fluctuating behaviour. It was originally thought that this complex behaviour was a result of the diameter of the mould colonies being approximately the same size as the width of the viscometer annulus, but since the width of the viscometer annulus was found to be

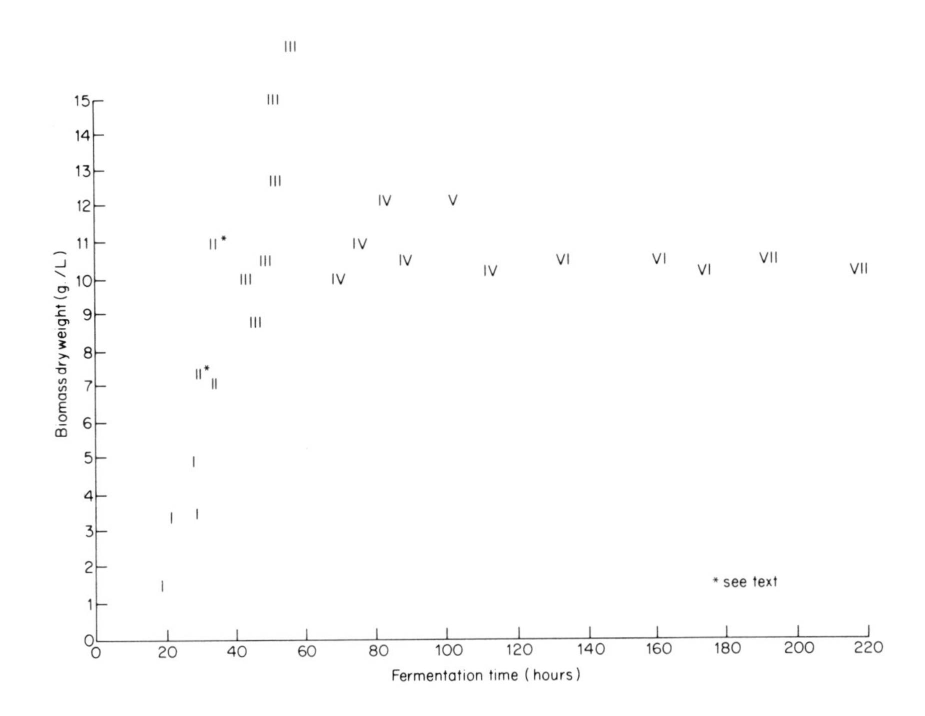

Fig. 9. Scatter diagram showing morphological groupings in relation to final biomass dry weight yields and time of fermentation to maximum biomass dry weight value.

approximately three times the colony-size, this explanation is discounted. The sudden increase in viscosity, followed by irregular fluctuations, suggests rather the presence of transient mechanical effects such as the interlacing of hyphae from adjacent colonies to form bridges or networks in the suspension (Fig. 9). Microscopic examination has on many occasions revealed the ability of Type I and Type III colonies to flow under the stimulus of gentle pressure on the cover-slip of wet-film preparations. This behaviour may also be observed directly in the diluted sample, for instance by agitating the suspension in a viewing-cell. It is noticeable even in the fermenter column that the suspension exhibits pronounced but localised streaming. The combination of particle flexibility with the ability to form transient matrices within the suspension may account for the unusual rheology of such broths, with ephemeral hyphal networks which repeatedly collapse in the face of excess shear to be replaced by streamlined particles offering minimal resistance to flow.

The importance of these findings for batch growth-kinetics may be judged by a comparison of the maximum biomass dry weights and the time taken to reach these maxima (Fig. 9) for a number of morphological types (Fig. 1,2,10,11). It can be seen that fermentations of Types IV-VII all achieve approximately the same dry weight figures, but vary in the time taken to reach this. Type III colonies attained the highest biomass-levels, although Types I and II had the highest growth-rates. The yields of Type II colonies were increased by the simple expedient of adding water to reduce the viscosity (denoted by II* in Fig. 9).

The biomass ceiling of around 12 $g/1^{-1}$ dry weight is much lower than levels commonly found in stirred ractors, where dry weights of 30-40 g 1^{-1} may be obtained.

Continuous culture-biomass retention

Continuous culture work represents a special case, in that the new colonies have to be formed from vegetative fragments of older colonies. Because of the configuration of the tower fermenter, fluidisation, flotation and sedimentation mechanisms begin to select for the retention of specific morphological types. Other types are removed by washout. Not only is there a gradiation of sedimentary forms within the column, but the effluent biomass appears different to that at the top of the fermenter, near to the swan-neck overflow.

The most significant feature of early continuous culture operation in 10 l capacity tower fermenters was that even during transient states, the culture resisted washout at hydraulic throughputs greater than the culture growth rate ($D > \mu$). In other words, the concentration of biomass in the effluent stream was always less than that in the fermenter itself.

The dependency of biomass-retention on morphological characteristics was highlighted on an occasion when, for some unknown reason, the organism formed extremely long filaments. The biomass concentration was much reduced from that expected. Instead, the fermenter behaved more like a conventional chemostat, the specific growth-rate corresponding to the dilution rate at steady-state.

Transient states on step-changes of dilution-rate

During transient states of continuous culture in the tower, there were considerable fluctuations in the biomass-

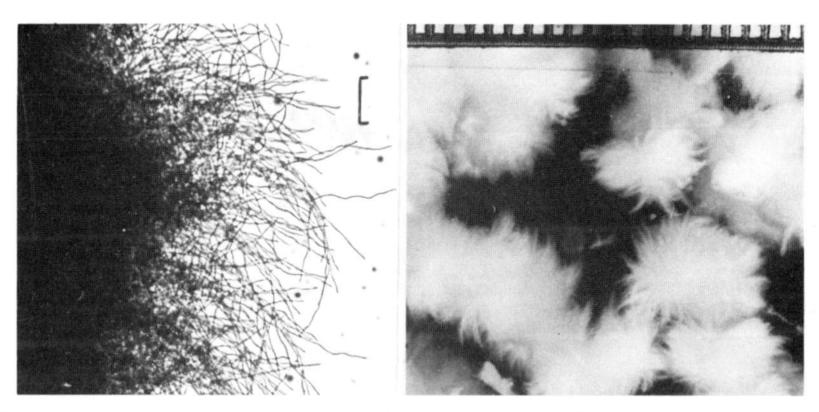

Fig. 10. Type IV colonies. (L.H. bright field - bar = 100 μm. R.H. actual appearance - 1 division = 1 mm.

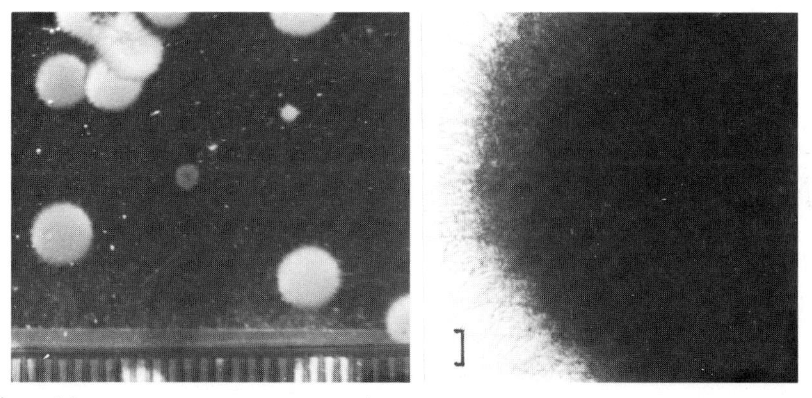

Fig. 10. Type V colonies. (L.H. actual appearance - 1 division = 1 mm. R.H. bright field - bar = 100 μm).

concentration and morphological appearance at each point along the column. The precise nature of this variation was apparently unpredictable.

Despite this, fermentations were very robust, with steady-states being held for several weeks, on one occasion culminating in a fermentation of 120 days duration. During such prolonged steady-states, it was obvious that colonies of mycelium gave rise to new colonies by a process of fragmentation and subsequent regrowth. The role of wall-growth was minimised

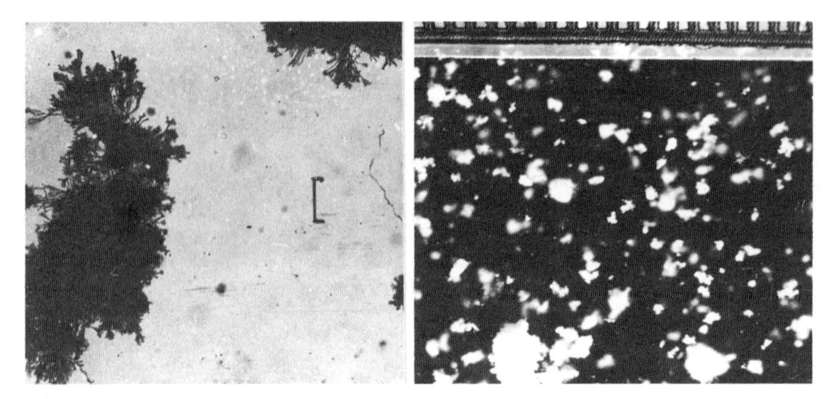

Fig. 11. Type V colonies. (L.H. actual appearance. 1 division = 1 mm; R.H. bright field - bar = 100 μm).

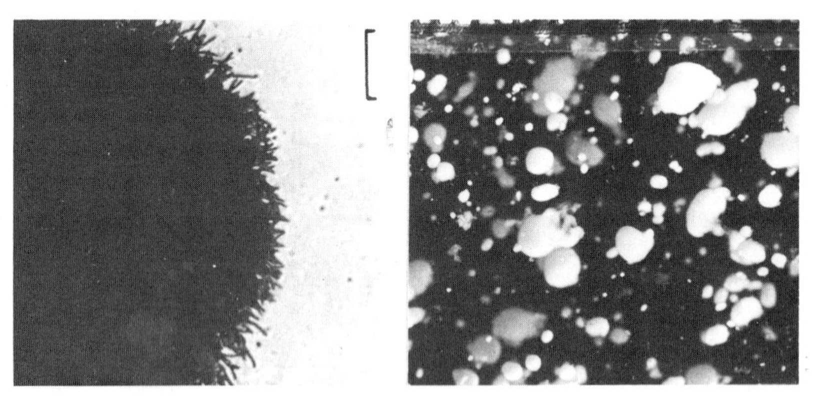

Fig. 11. Type VI colonies. (L.H. actual appearance. 1 division = 1 mm; R.H. bright field - bar = 100 μm).

---

by the smoothness of the vessel internals, the only sites of attachment being side the side-arms, and at the PTFE sparge-plate. It was felt unlikely that such growth had played any significant part in determining the behaviour of the fermenter.

High hydraulic feed-rates

Following the discovery that biomass-retention was a feature of the aerobic tower-fermenter, much as has been demonstrated in alcoholic fermentations with yeast (Shore & Royston, 1968), experiments were performed at very high hydraulic loadings. It was found that even though the maximum specific growth-rate remained at 0.26 h^{-1}, dilution-rates up to 7 h^{-1} were possible, although at such loadings, washout was almost complete and biomass levels very low (Pannell, 1976). The high value found for specific growth-rate may be attributable not only to a selection against slower-growing species of mycelium, but also to a selective removal by fluidisation/flotation mechanisms of vacuolated, presumably senescent, cells.

APPLICATIONS AND FUTURE PROSPECTS

Even the brief consideration given here reveals the use of a simple vertical column fermenter to be a more complex matter than first meets the eye.

Under steady-state conditions, the fermenter behaves as a three-phase fluidised bed, the engineering properties of which require considerable research. Commercial developments regarding the treatment of dilute effluents have already reached pilot-scale evaluation, indicating that in specialised applications, this type of fermentation system has its attractions. Possibly the most severe limitation in commercial applications is the biomass-level which can be achieved. Recent attempts to augment the sedimentary properties of biomass in column fermenters and to impose stricter regularities on the fluidisation behaviour have involved the use of inert support particles (Atkinson, 1974) and flocculating agents such as alginates (White & Portno, 1978), and it seems likely that such developments will lead to further commercial exploitation of such systems.

From the viewpoint of fundamental scientific investigation, the system of interaction between organism and fermentation environment in conditions of low shear offers information complementary to that obtainable in other systems such as stirred tanks. Where mechanical damage is a problem (Asai *et al.*, 1978) this feature may be particularly useful in a practical sense.

ACKNOWLEDGEMENT

The author wishes to acknowledge the contribution of Dr. S.D. Pannell, who, although unable to complete his co-authorship of this paper, was responsible for the continuous fermentation work.

REFERENCES

AIBA, S., KITAI, S. & ISHIDA, N. (1962). Density of yeast cells and viscosity of its suspension. *Journal of General Microbiology* 8, 103.

ASAI, T., SAWADA, H., YAMAGUCHI, T., SUZUKI, M., HIGASHIDE, E. & UCHIDA, M. (1978). The effect of mechanical damage on the production of Maridomycin by *Streptocomyces hygroscopicus*. *Journal of Fermentation Technology* 56 (4) 374-379.

ATKINSON, B. (1974). The principal types of fermenter. In *Biochemical Reactors* Published by Pion Limited, London.

EINSTEIN, A. (1911). Berischtigung 2u Meine Arbeit; 'Eine Neue Bestimmung Der Molekuldimensionen'. *Annal Physik* 34, 591.

JAMES, A. (1973). *PhD Thesis* University of Aston in Birmingham.

PANNEL, S.D. (1976). *PhD Thesis* University of Aston in Birmingham.

RYU, D.D.Y. & HUMPHREY, A.E. (1973). Examples of computer-aided fermentation systems. *Journal of Applied Chemistry and Biotechnology* 23 (4) 283-295.

SHORE, D.T. & ROYSTON, M.G. (1968). Chemical engineering of the continuous brewing process. *Chemical Engineer, London* 218 CE99-CE109.

WHITE, F.H. & PORTNO, A.D. (1978). Continuous fermentation by immobilised brewer's yeast. *Journal of Institute of Brewing* 84 (4) 228-230.

NOVEL DEVELOPMENTS IN MICROBIAL FILM REACTORS

J.G. Anderson[1] and J.A. Blain[2]

[1]*Department of Applied Microbiology.*
[2]*Department of Biochemistry,*
Strathclyde University, Glasgow, Scotland.

INTRODUCTION

Although the stirred tank reactor continues to be the most widely used type of fermenter the last decade has seen the emergence of a wide range of fermenter configurations developed for specific applications. The most notable trend in this development has been towards lowering investment and operating costs and this has led to the emergence of fermenter systems of unconventional design, mostly directed towards simpler construction and more efficient oxygen transfer (Kristiansen, 1978). Particularly favoured at the present time are various types of tubular reactors such as the pressure cycle fermenter and various types of bubble column and loop fermenters with either mechanical or pneumatic recycling. A detailed discussion of the design and performance of these systems has been presented by Katinger (1977).

For many conventional processes these tubular loop systems appear to offer ideal or near ideal solutions. There are, however, a range of more unusual applications for fermenters where the creation of more heterogeneous conditions is appropriate. In conventional fermenters the microbial population is freely suspended in the turbulent broth. An alternative arrangement is to make use of the natural capacity of microorganisms to attach to and develop as films on supporting solid surfaces. These biological film systems have had a longstanding although extremely limited range of applications. Recently, however, there has been much interest in the further exploitation of the attached mode of growth.

It would be inappropriate to discuss these biological film systems without reference to the term 'immobilised cell reactor' which is now being widely used in the field of biotechnology.

The concept of whole cell immobilisation has not been limited to procedures such as entrapment in polymeric matrices or adsorption of cells to ion exchange resins. Whole cell immobilisation has been defined as a physical confinement or localisation of microorganisms that permits their economical reuse (Abbott, 1977). Clearly, microbial film reactors, if appropriately applied, readily fall within the confines of this definition.

There are a number of potential advantages which stem from retaining cells in the reactor vessel during continuous processing. These advantages which have been fully described in previous reviews (Atkinson & Fowler, 1974; Abbott, 1977; Durand & Navarrow, 1978; Ash, 1979) include prevention of cell wash out, separation of cell production and biocatalysis processes, easier product recovery, reactor stability under varying substrate feed rates and high substrate conversion. Microbial film reactors can be operated to exploit these advantages and it is the wider appreciation of this potential which has resulted in the upsurge of interest in this type of system.

Apart from commercial considerations there is also a great need for more basic information on the nature of the interactions between microorganisms and surfaces not only in relation to fermentation (Ash, 1979) but also because of the importance of these phenomena in natural environments (Meadows & Anderson, 1979; Burns, 1979), and in host pathogen interactions (Arbuthnott & Smyth, 1979; Rutter, 1979). Studies on microbial-surface interactions can be conveniently carried out using special fermenters or alternative equipment designed to encourage film growth. Consequently the discussion which follows will deal not only with commercial applications of film reactors but also where such systems have been used for experimental studies on film-forming microorganisms. Although applications involving filamentous fungi will form the main subject of this discussion, brief coverage will also be given of the range of applications of microbial film reactors to indicate the current level of interest in these systems.

APPLICATION OF MICROBIAL FILMS

Packed bed film reactors

The traditional use of reactors in which the microbial population develops as a film on supporting surfaces has been in waste treatment plants and for the production of vinegar by the 'quick vinegar' process.

With the percolating filter (trickling filter) system for effluent treatment a mixed microbial population develops as a slime layer on the surfaces of packing material which is arranged as a bed. The packing media can be either conventional stoneware or, more recently, developed plastic media which have the advantage of lightness, high surface area and high void (interstitial) spaces. A relatively thick microbial film develops which is self regulating by sloughing and by macro-invertebrate grazing. In this system the film thickness is not accurately controlled and this has contributed to the difficulty of predicting accurately the performance of these reactors.

The major advantages of trickling filters compared with other effluent treatment systems is that no power is consumed for agitation or gas transfer so that operating costs are low. Also because of the heterogeneous nature of film growth these reactors can withstand shock loads of toxic materials more readily than completely mixed systems. A recent review of the design and application of waste water treatment systems, including percolating filters, has been presented by Winkler & Thomas (1977).

Information on the microbial populations of trickling filters is limited. Although it has been generally assumed that mycelial fungi are only of minor importance, studies by Cooke (1976) indicate that this may not be the case. Direct observations made on the slime layer attached to individual stones or to glass slides placed in the filter bed showed that fungal mycelia were an important structural and presumably metabolically active component of the attached population. Much more information is needed on the special groups of fungi and their physiological activities in trickling filters and other waste water treatment systems.

Although packed bed film reactors have traditionally been used as a means of aerobic effluent treatment, an anaerobic version (the anaerobic filter) has been developed which shows considerable promise for the economic treatment of dilute to medium strength waste waters (Anderson & Donnelly, 1977).

The packed bed concept also appears appropriate for certain bacterial leaching processes particularly the heap leaching of low grade sulphide bearing ores which is mediated by *Thiobacillus* species. The importance of cellular adhesion in this application has recently been discussed by Ash (1979). Apparently for the oxidation of elemental sulphur by *T. thiooxidans*, microbial attachment is a prerequisite for attack of the substrate. Also, attachment of the bacteria to the ore in heap leaching promotes a high concentration of bacteria in the reaction zone such as that the system behaves as a plug flow film reactor.

Rotating disc systems

Another biological film system developed for waste water treatment is the so called rotating biological contactor. In these units a complex microbial slime layer is built up on a series of partly submerged discs which rotate slowly in the vertical plane in a trough of waste water. As the discs turn the microbial film is exposed in succession to nutrient solution and to air. The film thickness is self regulating since excess biomass sloughs from the disc surface.

The rotating biological contactor was suggested many years ago but initially found little use. More recently however it has received much attention and is now being increasingly exploited for the treatment of both domestic and industrial effluents. Many new design features have been introduced and for some applications rotating discs have been replaced by honeycombed plastic elements which provide greater surface area for film growth. The major advantages of the system are the ability to resist shock loadings, short retention time, low power requirement, low sludge production and excellent process control (Borchardt, 1971). A comprehensive description of rotating biological contactors and of recent innovations and applications has been presented by Antonie (1975).

Recently attention has been drawn to the technical and economic potential of the rotating biological contactor for the biological treatment of acid mine drainage waters which are a serious pollution hazard to receiving streams (Olem & Unz, 1977). Biological treatment involves utilisation of bacteria of the species *Thiobacillus ferrooxidans* which efficiently transform ferrous iron to the ferric state. Detailed experiments have been carried out with small scale rotating contactors to investigate the major factors influencing chemolithotrophic iron oxidation by these organisms (Unz *et al.*, 1979).

In a totally different context a rotating disc method has also found use as a convenient experimental system for studying the adhesion of oral streptococci to solid surfaces (Rutter & Abbott, 1978). Various types of equipment have been used for studies on plaque forming bacteria and the particular advantage claimed for the rotating disc method was an account of the well defined hydrodynamics associated with disc rotation in liquid medium.

Rotating tube or drum systems

Various types of equipment have been described in which the microbial film is supported on a rotating drum or tube. These include the rotary spiral fermenter (Ugolini, 1960), the rotating tube reactor (Tomlinson & Snaddon, 1966) and the concentric cylinder reactor (Kornegay & Andrews, 1968). These systems have been used mainly for experimental studies on the development and biological activity of microbial films. Less elaborate apparatus comprising simple plates of various materials have also been used for these purposes. A comprehensive discussion of the equipment employed and results obtained in these studies has been presented in a review by Atkinson & Fowler (1974).

Mixed microbial film fermenters

The use of small particles in a stirred tank or fluidised bed fermenter creates a large surface area for microbial film development. Abrasion resulting from physical contact between the moving particles acts to remove excess surface growth and maintain a relatively constant film thickness. This concept has been developed and studied extensively by Atkinson and co-workers. The design features of these systems, together with studies on their performance, particularly in relation to oxygen and substrate diffusion within the films, have been discussed in a number of important reviews (Atkinson, 1973, 1974; Atkinson & Fowler, 1974; Atkinson & Daoud, 1976). An important feature of the design is the control obtained over film thickness. Because of this both operational control and prediction of the performance of the reactor are considerably simplified.

In the fluidised bed system sedimentation forces act to maintain the biomass coated particles in the reactor at relatively high liquid throughput rates. Incorporation of liquid recycle into this system generates very uniform conditions throughout the reactor and creates what has been called the completely mixed microbial film fermenter or CMMFF (Atkinson & Knights, 1975).

Recently the CMMFF concept has been applied with modification for the cultivation of mycelial fungi (Atkinson & Lewis - this Volume). In this system particles of compressed woven stainless steel mesh are used to support attached growth of *Aspergillus foetidus*. Mycelium grows within each particle producing a 'floc' which is kept a uniform size since overgrowth is abraded from the external surfaces.

The system to some extent shares certain features associated with the more conventional pelleted growth form. It has for example been reported that simple towers operated with pelleted filamentous fungi can make use of colony sedimentation to achieve biomass retention during continuous medium throughput (Cocker & Greenshields, 1977). This feature depends on the poor vertical mixing and consequent heterogeneity which occurs in simple tower systems. Recently, however, it has been reported that it is also possible to obtain selective biomass retention with filamentous fungi in a well-mixed air lift recirculating tower fermenter by careful location and design of the effluent take-off (Kristiansen & Bu'Lock, 1980). Reference to these applications of the tower system has been made here to illustrate that biomass retention of mycelial fungi is also possible by techniques other than film growth. A totally different type of particle film fermenter to the CMMFF design has been described by Englebart & Englebart (1975). This system also seeks to combine the advantages of adhesive and submerged cultivation. Rod or filament shaped particles are used in the reactor to provide a large surface area for adhering organisms. The reactor also contains a large number of tubes of several centimetres in diameter each having an internal helical core element. The particles have a specific gravity sufficiently different from that of the medium so that the liquid can be passed through the tubes at a very fast rate relative to the suspended film carrying particles. The possibility of creating relative speeds of 10-100 cm/sec between the microbial film and the medium are claimed by this arrangement. With conventional fermenters even very turbulent conditions produce only low relative speeds between the suspended cells and the surrounding medium. The principal advantage of generating these high relative speeds is to increase diffusion speeds and consequently accelerate exchange of substances between the adhering organisms and the liquid stream.

FILM FORMATION BY MYCELIAL FUNGI

The ability of filamentous fungi to attach to and colonise immersed solid surfaces has attracted attention mainly as a nuisance characteristic and potential source of problems in fermentation processes which utilise these organisms. Wall growth and accumulations of adhering mycelium around medium inlets and overflow pipes and on sensing probes are all problems familiar to those who work with fermenter cultures of filamentous fungi.

Fouling problems can be particularly acute when operating under continuous flow conditions. Here the ecological advantages of attachment are evident since the organism can maintain a resident biomass even under conditions which would result in rapid wash out of a non-attached population. However, for the operator such behaviour can lead to serious problems. In small vessels accumulations of attached growth decrease the culture volume and create heterogeneous culture conditions. Periodic blockages of the medium inlet result in interruption of the nutrient supply leading to transient changes in culture physiology. Similarly, complete or partial blockage of the medium overflow results in biomass retention with resultant dramatic changes in growth kinetics. Solutions to these problems which have been discussed by Bull & Bushell (1976) include appropriate positioning of the overflow pipe, the use of materials such as teflon to discourage attachment and the use of periodic changes in vessel pressure and impeller speed to physically remove attached growth.

Problems arising from surface growths of fungi also arise in industrial scale operations. Apart from the more obvious wall, impeller and port fouling problems, attached growth can encourage corrosion and can lead to a deterioration in performance of heat exchangers and other equipment downstream of the fermenter (Ash, 1979). Fouling problems can be so acute as to discourage the use of strains which are excessively adherent.

Problems associated with attached growth of filamentous fungi are not restricted to fermentation work. Many filamentous fungi are important biodeteriogens and it is likely, particularly in aquatic situations, that the attachment of spores or hyphal fragments represents an important initial stage in the biodeterioration process.

Although the attachment of bacteria to solid surfaces has been extensively studied there appears to have been no equivalent work carried out with fungi. It has been suggested that surface growth of bacteria in fermenters may result from an electrostatic attraction between the organism and the surfaces involved (Munson & Bridges, 1964). It is, however, extremely unlikely that any single mechanism can explain surface attachment. The adhesion of microorganisms to surfaces is now viewed as a complex multi-stage process involving interaction between the microbial surface and a range of physical forces (Ellwood *et al.*, 1979; Curtis, 1979).

The filamentous nature of mycelial fungi gives these organisms unique properties which appear well suited to colonial growth on surfaces under turbulent liquid conditions. A good deal is now known about the mechanisms of pellet formation by filamentous fungi in deep culture (Whitaker & Long, 1973; Atkinson & Daoud, 1976; Metz & Kossen, 1977). This type of colonial development also results from a complex interaction between physiological and cultural conditions and it is highly likely that some of the factors involved here, particularly those influencing spore and hyphal agglomeration, will also be important for the attachment and subsequent development of fungi at solid state surfaces. Clearly studies on fungal attachment could provide useful information not only in relation to the industrial exploitation of fungi but also for a better understanding of the ecology of these organisms in aquatic environments.

THE DISC FERMENTER

Apart from the work at Atkinson and co-workers with the fluidised bed fermenter, there appears to have been little, if any, attempt to deliberately utilise the natural tendency of filamentous fungi to attach to and colonise immersed solid surfaces. The disc fermenter described here has been designed specifically for this purpose. It can be used as a convenient experimental system for studying attachment properties of mycelial fungi and for examining possible applications of the attached growth mode. The fermenter, which utilises the rotating disc principle, facilitates solid surface cultivation under highly controlled sterile conditions and the design permits aseptic sampling of both culture fluid and attached biomass.

The design features of a small scale disc fermenter are illustrated in Fig. 1. The fermenter consists of a standard QVF T-piece, normally 30.5 x 16.2 cm with a 10.2 cm side arm. Stainless steel end plates are clamped to the glass vessel with standard QVF flanges. There are ports in the end plates for fluid sampling and filling and for the passage of filtered sterile air over the discs. A 9 mm diameter stainless steel shaft carried from 9 to 18 discs of 13.3 cm diameter and 4.5 mm thickness, made of roughened high density polypropylene. The discs are separated by plastic or stainless steel washers and these are tightened together by knurled screws at each end of the shaft. Disc spacing can be varied to suit the growth characteristics of the organisms.

At the drive end the shaft passes through a sterile seal in the end plate. The shaft is driven by a geared motor with variable speed control. Generally a low disc rotational speed (9 r.p.m.) has been used for cultivation of filamentous fungi.

An important consideration when designing the fermenter was the need to obtain representative samples of attached mycelium for estimation of growth rates and for analytical studies on mycelial composition and metabolic activities. Because of the filamentous felt-like nature of the attached mycelial layer, it was not possible to remove portions from the disc surfaces by the use of, for example, sampling tabs which have proved adequate for sampling bacterial films on rotating contactors (Unz *et al.*, 1979). Instead, special modified discs were designed and three or four of these were located on the shaft directly under the QVF side arm (Fig. 1). Each sampling disc consisted of four detachable segments which had, on their inner surface, two apertures which fitted over two stainless steel projections rising from a collar on the shaft.

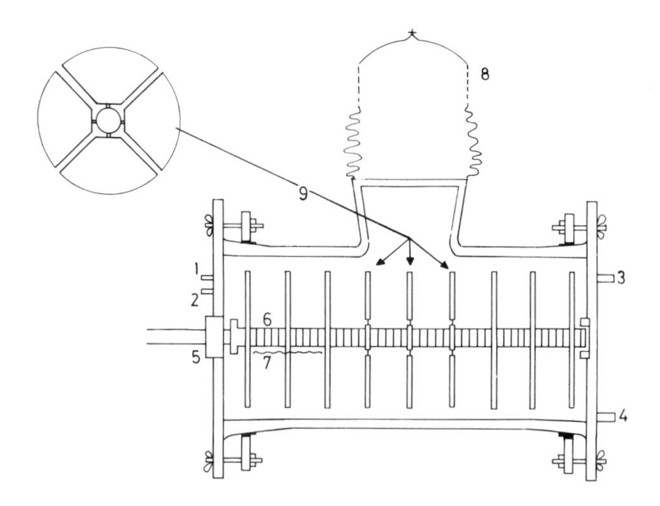

Fig. 1. Design features of the disc fermenter (Blain *et al.*, 1979). (1) Air inlet. (2) Inoculation port. (3) Air outlet. (4) Sampling port. (5) Shaft seal. (6) Spacers. (7) Medium level. (8) Sampling bag. (9) Detachable sampling segments.

The procedure which was initially used to obtain aseptic access into the fermenter for the removal of disc segments is also indicated in Fig. 1. A long autoclavable plastic bag was pushed into the side arm and a detachable segment with adherent biomass was gripped and withdrawn. The top of the bag was then pinched off and two seals were made approximately 1-2 cm apart. The plastic between the two seals was then cut so that the portion of the bag containing the sample was removed and a sealed, although a slightly shorter bag remained on the fermenter for further sampling.

Subsequently a more robust device was designed for the aseptic removal of disc segments (Fig. 2). This device consisted of a short polypropylene cylinder clamped to the fermenter side arm. An aperture of 6 cm diameter on the side of the cylinder had internal and external threaded cylindrical projections. A polypropylene bottle was screwed onto the outer projection and a polypropylene cap was screwed onto the inner projection. A glove-box type silicone rubber glove fitted on the top end of the cylinder enabled the internal cap to be removed and left hanging by a supporting cord while a removable segment was taken from the disc assembly and placed in the bottle. The internal cap was then replaced before removing the bottle.

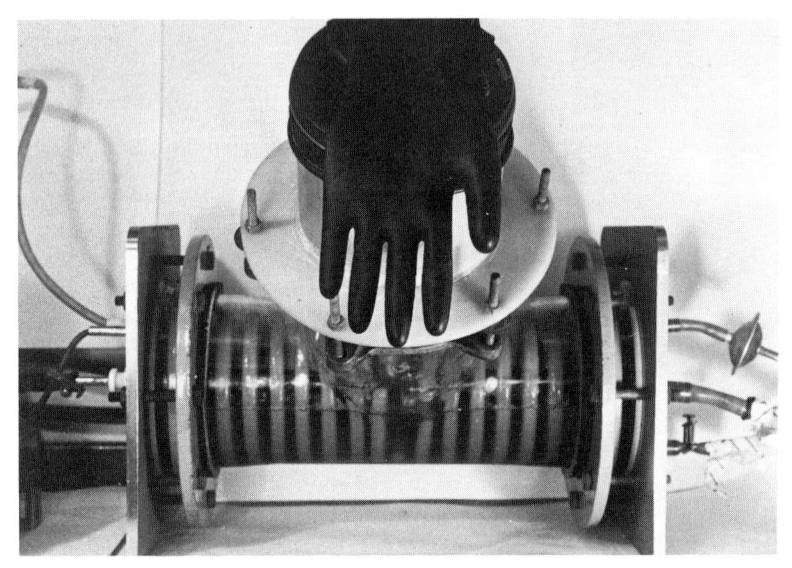

Fig. 2. 'Glovebox' device fitted to the glass vessel side-arm of the disc fermenter for aseptic removal of disc segments.

A larger scale disc fermenter has also been constructed with a 40 l working volume (Fig. 3). This vessel does not have the facility for periodic biomass sampling. Removal and collection of the complete biomass at the end of a batch operation could however be readily carried out. To allow for biomass removal a small portion of the outer curved edge of each disc was removed leaving a flat area. During assembly of the fermenter the flat edge portions of the discs are lined up. After collection of fluid from the fermenter the discs are turned so that the flat edges are situated at the top of the fermenter. A long thin tube is then inserted through the end plate between the flat edge of the discs and the glass vessel. The tube has a downward pointing nozzle through which a high pressure water jet can be directed at the mycelial layers which peel from the disc surfaces and collect along the bottom of the glass vessel. After

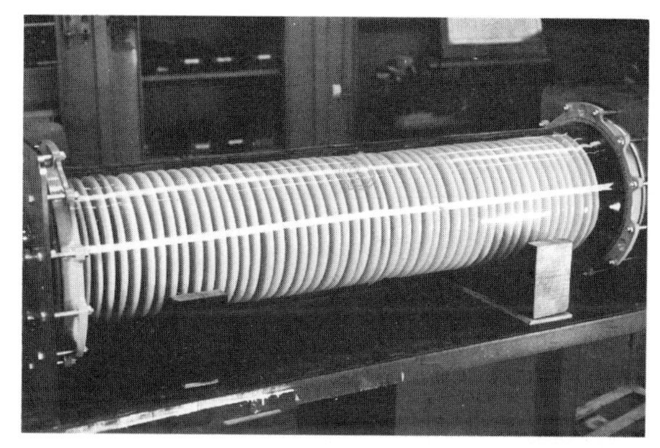

Fig. 3. Disc fermenter (40 litre working volume) showing series of polypropylene discs prior to inoculation and film growth.

withdrawing the tube the discs are again turned so that the flat edges are positioned in the bottom of the vessel. This allows sufficient space between the discs and the glass for the biomass to be removed through a broad port either by inclining the vessel or by a push rod device.

Environmental conditions in this scaled up version can be monitored by recycling the medium through an external circuit in which sensing probes are located (Fig. 4.).

Inoculation and film growth

Dormant or pregerminated spores are used to inoculate the fermenter and these attach to the disc surfaces and initiate film growth. Inoculation of mycelium particularly if aggregated or pelleted, has proved less suitable since all of the inoculum may not attach to the discs and consequently some development also occurs in the bulk medium.

During growth the film thickens to form a spongy layer which holds an increasing proportion of the medium so that the level in the vessel tends to decrease. This is not a problem when continuous liquid feed is employed, however, with batch operation further introduction of fluid is required to maintain the bulk medium level.

The system has been tested with a wide range of filamentous fungi including representative of the genera *Aspergillus*, *Rhizopus*, *Mucor* and *Penicillium* all of which grew successfully on the discs. Normally the disc surfaces are rapidly

Fig. 4. Disc fermenter (40 litre) showing discs coated with *A. niger* mycelia and in foreground, electrode assembly situated in external loop for medium recirculation.

colonised by hyphae to form a thin even film which then thickens as growth proceeds. However, with one organism, an *A. terreus* mutant, numerous discrete domed colonies developed on the discs. This strain lacked the radial spreading ability normally found with filamentous fungi.

Sporulation does not occur on the disc surfaces while rotational submergence is maintained. Removal of the medium can however result in rapid 'synchronised' sporulation on the disc surfaces. It is possible that this technique might be of use for large scale spore production since not only is there a high surface area for sporulation but also the complete process, including spore harvesting, can be carried out under highly aseptic conditions.

With *A. niger*, which has been most extensively studied, the mycelial layer normally remains 'stable' without detachment even over prolonged cultivation periods of up to several weeks. Certain media formulations however, which cause extensive autolysis in deep culture, can also cause breakdown and detachment of the mycelial layers in the disc fermenter.

Oxygen transfer considerations

Characterisation of oxygen transfer processes in culture systems which employ colonial growth forms of microorganisms has proved extremely complex. Although a great deal of work has been done with pellet cultures of mycelial fungi there is still considerable controversy over the relative importance of the various factors which influence the rates of oxygen uptake into pellets (Miura, 1976; Metz & Kossen, 1977; Tsao & Lee, 1977). With biological film systems the work of Atkinson and co-workers (see Atkinson & Fowler, 1974) on modelling substrate diffusion and consumption in films has been particularly notable. It is not yet clear however the extent to which these models can be applied to different microbial film systems which may be exposed to quite different physical conditions.

With the disc fermenter initial studies were carried out to characterise the oxygen transfer capacity of the fermenter in the absence of the microbial population (Artavanis & Todd, 1980). These studies attempted to ascertain K_La values for the copper catalysed oxidation of sodium sulphite. However, application of the sulphite method to the disc fermenter produced some apparent discrepancies which were subsequently shown to be due to two incorrect assumptions under the prevailing mass transfer conditions.

Sulphite oxidation in the disc fermenter was characterised by a high rate over the initial 10-15 min followed by a decline to a constant rate after some 40 min. The peak rate was approximately twelve times higher than the final constant rate. This pattern was noticed over the complete range of conditions studied including disc rotational speed and fractional submergence in the liquid. Thus variation of factors, expected to influence the rate of oxygen diffusion, such as liquid film thickness, residence time of the liquid film in the gas phase and agitation intensity in the bulk liquid, did not affect the final constant rate of oxygen uptake. It was concluded that, under the prevailing mass transfer conditions in the disc fermenter, the sulphite oxidation reaction was not diffusion controlled and therefore could not be applied for measurement of true physical mass transfer coefficients in the disc fermenter.

Artavanis & Todd (1980) were however able to estimate a K_L value for the disc fermenter based on knowledge of the final constant rate of sulphite oxidation and of the interfacial area of the liquid film. Thus based on a final steady state oxygen uptake rate of 0.125 millimols/l min and a wetted interfacial area of 127 m^2/m^3 (liquid volume 2.0 l, disc submergence 35% of total area) K_L was calculated to be 8.46×10^{-5} m/s for 0.25 M sodium sulphite. This was similar to K_L values previously determined with small-scale rotating disc assemblies.

Although measurement of mass transfer coefficients in fermenters are often made for convenience in non-biological conditions, the information provided is extremely limited, particularly for processes in which biological films or mycelial organisms are to be employed. With the disc fermenter growth of *A. niger* in continuous feed conditions or in batch runs employing high substrate concentrations could result in the development of spongy layers of from 5-10 mm thickness within 72 h. Clearly under these conditions the mass transfer characteristics of the system will be radically altered. Also, prime consideration must be given to the extent of oxygen and substrate limitation within the mycelial layer.

Several studies have been carried out to investigate the consumption of oxygen and substrate by growing microbial films (Tomlinson & Snaddon, 1966; Kornegay & Andrews, 1968). The results obtained in these studies indicated that only the outer 70-100 μm of the film was 'active'. Although the films increased in thickness beyond these values, the rate of substrate consumption maintained a constant value.

An oxygen micro-probe has also been used to determine the dissolved oxygen profile next to and within a biological film (Bungay *et al.*, 1969). These results also indicated that 'active' respiratory metabolism was only possible in the outermost region of the film. The depth of the aerobic layer, as determined by microprobe insertion, was found to be from 50-150 μm depending on the substrate concentration supplied to the film.

Atkinson & Fowler (1974) have drawn attention to the good measure of agreement obtained in these different studies regarding the thickness of the 'active' layer. The microbial films examined in these studies were 'natural' films composed of mixed microbial species and may well have been predominantly bacterial. Purely fungal films may possess somewhat different properties associated with, for example, different cell packing densities. Studies in progress on the spongy *A. niger* mycelial layers produced in the disc fermenter have shown that the rate of substrate uptake continued to increase with increases well in excess of 150 μm.

In terms of structure and density the mycelial layer produced in the disc fermenter is more likely to resemble a fungal pellet than a bacterial film. Currently there is much interest in the characterisation of oxygen and substrate transfer in mycelial pellets.

In a number of studies it has been assumed that the pellet behaves as a rigid but porous sphere and that molecular diffusion is the controlling mechanism for oxygen penetration. For situations in which oxygen can enter only by simple diffusion, Pirt (1966) has calculated that growth of the pellet would be restricted to a peripheral zone 77 μm wide. This prediction appears to be substantiated by Phillips (1966) data on *Penicillium chrysogenum* pellets.

In a more recent study Huang & Bungay (1973) have used micro-electrodes to measure the oxygen concentration in pellets of *A. niger*. In this study no oxygen was detected at a depth of about 135 μm into the pellet. The oxygen diffusivity within pellets reported in this study was 2.9 x 10^{-6} cm^2/sec.

Miura (1976) has however pointed out that these measurements were carried out for a fixed pellet in a slow flowing medium (28.3 cm/min) and that this situation may be very different from that in agitated culture conditions. To evaluate the influence of agitation and turbulence on mass transfer Miura examined the penetration rate of blue dextran into shaken pellets of *A. niger*. It was found that the mass transfer rate of the blue dextran, which was independent of physiological reactions, was highly influenced by agitation. Miura concluded that the effective diffusion coefficient of oxygen in the agitated mycelial pellet was of the order of 10^{-4} cm^2/sec and that this was greater by about one order of magnitude than the molecular diffusion coefficient. Increased oxygen uptake resulting from agitation was considered to be caused by medium penetration into pellets by turbulence and also by pellet deformations resulting in fluid exchange. These factors are likely to be of even greater significance in rapidly stirred cultures where velocity fluctuations near the impeller can be so severe as to cause outright rupture of pellets (Taguchi, 1971).

Indirect evidence on the importance of mass transfer of oxygen into pellets has also been given by Trinci (1970) in a study of the growth kinetics of individual shaken pellets of *A. terreus*. From the growth kinetic data it was calculated that the width of the peripheral growth zone of the pellet was of the order of several mm. Since this far exceeded the width of a growth zone which could rely only on molecular diffusion it was concluded that the texture of the pellet must have been sufficiently loose to permit medium penetration by turbulence and consequently mass transport of oxygen.

In the disc fermenter the slowly rotating mycelial layer is not subjected to the turbulent conditions experienced by pellets in shaken or stirred culture. However, the layer as it rotates above the bulk medium is subjected to gravitational force which results in fluid exchange. With thick layers this feature can be observed visually by stopping disc rotation. When stopped, the spongy layer on the upper exposed portion of each disc slowly collapses as the fluid held in the layer drains into the bulk medium. At the slow rotational speeds used (normally 8 r.p.m.) this mechanism will operate to some extent continuously. Fluid lost as the layer rotates above the medium is then replaced by medium uptake during rotational submergence.

An indication of the substantial exchange which takes

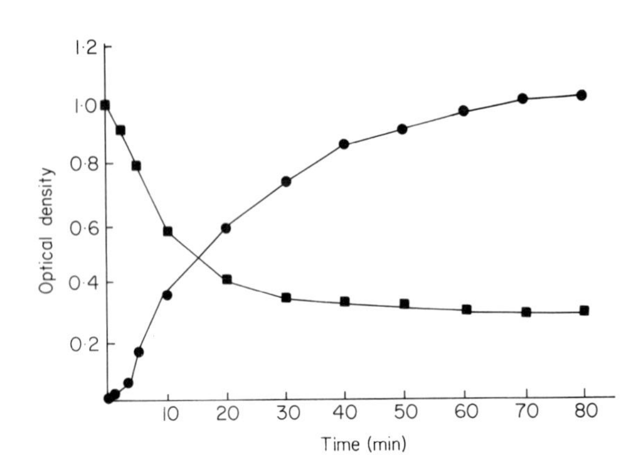

Fig. 5. Rates of dye uptake (●) and release (■) for *A. niger* mycelia attached to rotating discs (Blain *et al.*, 1979).

place between the bulk medium and fluid held in the rotating mycelial layer is shown in Fig. 5. In this experiment *A. niger* was grown on a high sucrose (citric acid production) medium for eight days, the medium was then removed and immediately replaced by a methylene blue solution. The discs were rotated at 8 r.p.m. and sample segments were removed at intervals. The dye concentration in the liquid held by the mycelial layer was measured at each sampling time. A similar experiment was carried out in which the mycelium was first soaked in dye solution and then the rate of dye release, from the mycelium into water, was measured. The degree of mixing indicated by these results was consistent with subsequent findings on the viability of the mycelium at different depths in the layer. For viability measurements portions of the biomass on sampled segments were removed and respiratory activities determined by Warburg manometry using medium removed from the fermenter at the time of sampling as substrate. The mycelium was removed as one complete layer after four days growth. After eleven days the thicker mycelial layer was separated into two portions comprising the inner and outer zones. After fifteen days the mycelial layer, which was approximately 9 mm thick, was separated into three portions comprising the inner, middle and outer zones. The respiratory activities of the mycelial samples obtained at different times and from different depths in the layer are shown in Table 1 along with C/N ratios for the samples.

The Warburg results obtained, although uninformative on respiratory rates under *in situ* conditions, did demonstrate

Table 1. Respiration rates and C/N ratios of *Aspergillus niger* mycelium from disc surfaces (from Blain *et al.*, 1979).

Age of mycelium (days)	Part of mycelial layer sampled	Oxygen uptake rate of mycelium (μl/mg/min)	C/N ratio of mycelium
4	Complete layer	0.56	7.75
11	Inner zone	0.47	10.3
	Outer zone	0.32	7.2
15	Inner zone	0.36	14.7
	Middle zone	0.44	8.6
	Outer zone	0.23	8.4

the continued viability of the mycelium throughout the depth of the layer. The composition of the mycelium within the layer was however heterogeneous as shown by differences in C/N ratios of the samples taken from different depths.

It is interesting to compare these results with a previous study by Yoshida *et al.* (1978) on respiratory activity within large (24 mm diameter) pellets of *Lentinus edodes*. In this study it was found that the Qo_2 of the mycelium decreased with distance from the circumference of the pellet and fell to zero for the central 8 mm portion. Cytological observations have also indicated the extent to which oxygen and substrate limitation can affect the viability of mycelium within mycelial pellets. Conditions in the centre of large pellets can be so severe that autolysis of the inner mycelium results in the formation of hollow pellets (Camici *et al.*, 1952; Clark, 1962; Yanagita & Kogane, 1963).

The viability of the innermost portion of the mycelium on the rotating discs may explain the continued adherence of the thick layer during prolonged cultivation periods of several weeks. This feature of disc cultivation was surprising in view of previous reported properties of biological film systems. Generally it has been observed that as biological films increase in thickness oxygen and substrate limitation in the deeper parts of the film result in an accumulation of gas and loss of adhesion between the film and the supporting surface so that the film falls away under the action of its own weight (Atkinson, 1974).

Much further work is needed in order to characterise the properties of fungal films; particularly the extent to which properties vary with different species and under different nutritional and physical growth conditions. Considering the extent to which the structure of pellets of a single species can vary under different growth conditions it is unlikely that any simple model could be used to predict, in general, the properties of fungal films.

Fermentation studies with the disc fermenter

While the heterogeneous conditions in the disc fermenter are not optimal for growth rate, they do offer a unique approach to the study of fungal product formation. Surface growth has been a traditional method of culture for some products and although this method still has limited application, much greater emphasis has been placed in recent years on the use of submerged batch culture. Since the disc mode of cultivation shares, to some extent, features of both surface and deep culture, it cannot be anticipated what effects will arise from the use of this 'intermediate' technique.

A complex interaction occurs between organism physiology and reactor design (Finn & Fiechter, 1979). This interaction is particularly significant with mycelial fungi which, because of their filamentous nature and plasticity of form can exhibit a wide range of cellular and colonial morphologies in response to the different physical and chemical conditions found in fermentation systems. It is highly likely that there are a number of potentially useful fungal products which are produced under the heterogeneous conditions found in solid or liquid surface colonies or within submerged film growth which might never be produced in the more homogeneous conditions imposed by submerged agitated conditions.

There can also be differential levels of product formation in different cultural conditions for reasons which are not readily apparent. Fig. 6 shows a result obtained during a comparison of the cultivation of *Lasiodiplodia theobromae* by liquid surface culture, shaken culture and disc culture techniques. Much higher levels of highly viscous mellein were produced in the disc fermenter than in surface or flask culture. The explanation for this may relate to physiological differences induced by growth form or alternatively to differences in typical fermenter parameters such as oxygen transfer. Nevertheless, the result serves to illustrate the frequently unpredictable effect mediated by a change in growth mode.

Studies have also been carried out using the disc fermenter to examine the effects of various parameters on the conversion of sucrose to citric acid by *A. niger* (Anderson *et al.*, 1980). Examination of this fermentation seemed particuarly appropriate in view of the wealth of comparative, although 'academic' literature available. Also, it has been well documented that the fermentation cycle is biphasic with an initial growth phase followed by a phase of growth limitation and product formation (Lockwood, 1975; Berry *et al.*, 1977). Production processes appear to be based on this consideration and best results have been obtained by altering the cultural conditions at appropriate times in order to optimise conditions for each stage (Smith *et al.*, 1974). The disc fermenter offers the facility of rapid medium replacement to an established biomass and consequently can be used to study optimal conditions for each phase.

An initial comparison was made between disc and conventional surface culture using a standard test strain of *A. niger* and a synthetic sucrose medium which was not replaced during the course of the fermentation. Under these conditions the disc system gave a higher yield in a shorter period although conversion rates were low in both systems (Fig. 7). In the disc fermenter mycelial growth continued during the fermentation to such an extent that in prolonged runs of from 8-15

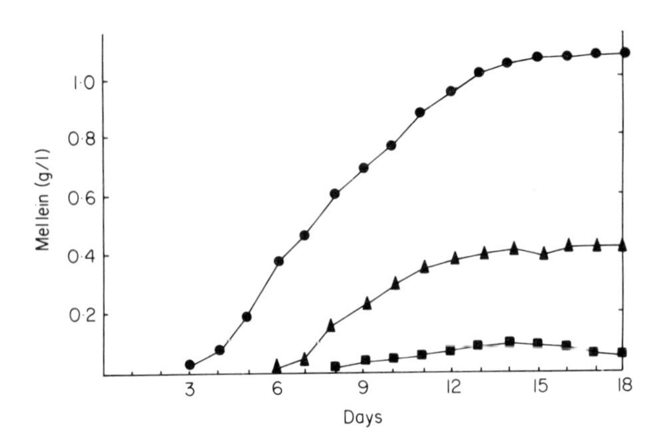

Fig. 6. Production of mellein 'gum' by *Lasiodiplodia theobromae* in surface (■—■), shaken (▲—▲), and disc fermenter (●—●) cultures.

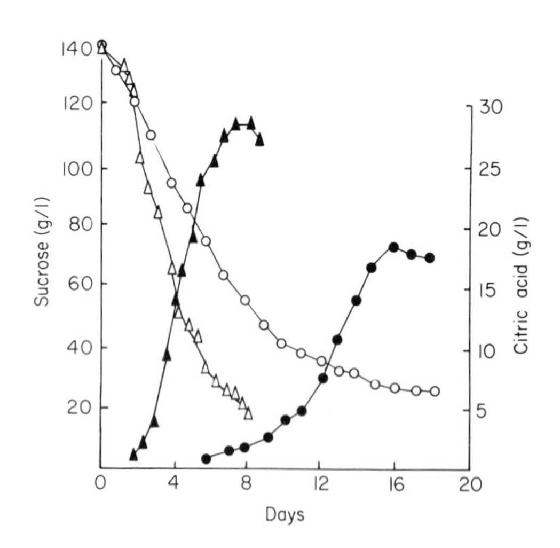

Fig. 7. Comparison of sucrose utilisation (△—△) and citric acid production (▲—▲) in disc culture with sucrose utilisation (○—○) and citric acid production (●—●) in surface culture.

days the inter-disc spaces became blocked. Consequently experiments were carried out in which the initial growth medium was replaced with media having growth-limiting deficiencies particularly in available nitrogen and phosphate. This restrained mycelial growth and left adequate inter-disc spaces for nutrient and oxygen transfer.

This replacement technique gave better yields than were obtained with uninterrupted (single medium) batch productions using conventional fermentation systems (Table 2).

Table 2. Comparison of the disc fermenter with conventional systems for the production of citric acid by *A. niger* (from Anderson *et al.*, 1980)

Fermentation System	Duration of fermentation (days)	Biomass produced (g/l)	Citric acid produced (g/l)
Surface	17	12	22
Shake flask	7	11	10
Stirred fermenter	7	19	9
Disc fermenter (using medium replacement)	15	8	80

It was observed that the increased citric acid yield obtained by rapid phosphate and nitrogen limitation was associated with low biomass production.

Further observations illustrating the use of medium replacement for studying citric acid production in the disc fermenter are given in Fig. 8. Results are given for a series of experiments in which the fermentation was initiated with a growth medium containing 14% sucrose which was then replaced after three days with sucrose solutions of differing concentrations. Before media changes acid production rates in all cases were about 0.3 g/l/h. The data show that production rates were diminished and remained low over the test period when the replacement solution contained 2% and 6% sucrose but were increased at sucrose concentrations of 10% and above, the highest ultimate rate being at 14%. This result was somewhat higher than was previously found by Chmiel (1977) for a different *A. niger* strain. From kinetic data obtained using a two stage fermentation process Chmiel found that the optimum sucrose concentration for citric acid production was in the range 30-100 g/l.

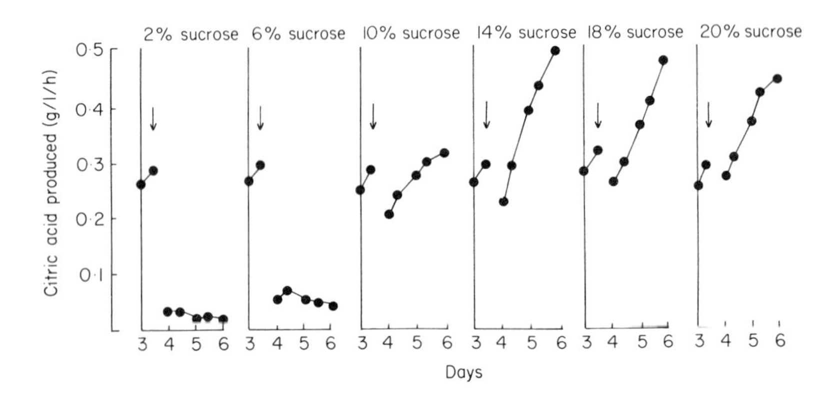

Fig. 8. Effect of sucrose concentration on rates of citric acid production in the disc fermenter with changes (arrowed) of sucrose concentration from 14% (Anderson *et al.*, 1980).

Chmiel (1977) also observed that citric acid accumulation above 10 g/l had an inhibitory effect on acid production. Metabolic inhibition by product accumulation is a factor of some significance in a number of processes employing conventional batch operation. Where this is particularly severe dialysis culture techniques may have application (Finn & Fiechter, 1979). An alternative approach could be provided by the use of film cultivation systems which overcome the problem by allowing periodic or continuous product removal without depletion of biomass.

This potential was examined in relation to citric acid production in the disc fermenter by supplying a continuous medium feed over an extended period. Mycelia were allowed to develop for three days on the initial growth medium which was then replaced by a 14% sucrose solution. A continuous feed of 14% sucrose was then established at a flow rate of 9 ml/h. The rate of citric acid production was measured before replacement and subsequently over a period of 27 days. During continuous feed the production rate increased to 0.57 g/l/h at the tenth day, after which it decreased slowly over the rest of the test period (Fig. 9).

Gradual loss of activity during continuous medium feed could not be ascribed to either substrate consumption (to non-optimal substrate levels) or to acid accumulation. The effect may be an inevitable consequence of cellular ageing. However, since the replacement solution contained only sucrose it is likely that further studies could define a more appropriate medium formulation which, although preventing growth, would

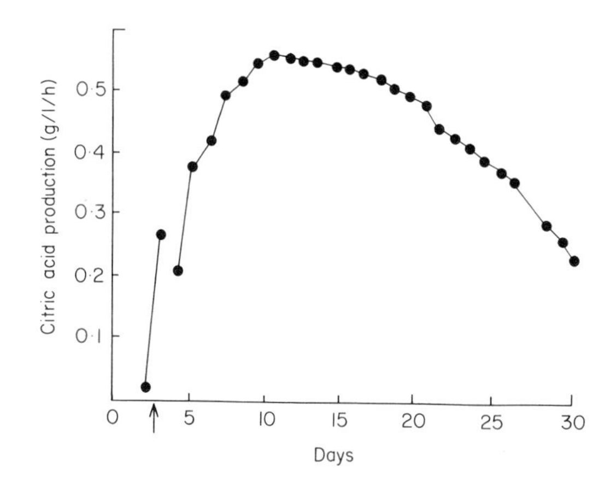

Fig. 9. Rates of citric acid production in the disc fermenter with medium exchange at 3 days and continuous feed from 8 days (Anderson *et al.*, 1980).

sustain a higher level of mycelial activity.

Clearly much further work is required to more fully explore the potential of film growth systems for special applications in fermentation. It should be borne in mind that the main objective of the work described here on citric acid production was to explore utilisation of the disc fermenter rather than to attempt to optimise citric acid production to competitive levels, an awesome task considering the massive research which has been required for the development of the present commercial process.

Currently there is much interest in the use of immobilised whole cells for substrate transformations. Most of the immobilised cell applications involving filamentous fungi have been concerned with carbohydrate transformations (see Chibata & Tosa, 1977). *A. niger* has been used for the production of gluconic acid from glucose by various immobilisation techniques including entrapment in polyacrylamide gels, flocculation by polyelectrolytes and binding gluteraldehyde treated cells to methacrylate polymers. An alternative approach to fungal 'immobilisation' has been the use of fungal pellets. A striking example is the use of mycelial pellets of *Mortierella vinacea* for the hydrolysis of raffinose in beet molasses (see Abbott, 1977). In this process agitated pellets are retained by fine mesh screens in a long multi-compartment tank through which beet molasses is passed. Excess raffinose in the molasses, which can adversely affect sucrose recovery, is hydrolysed to sucrose and galactose by α-galactosidase in the pelleted mycelia.

For some such applications it might be advantageous to utilise a film reactor, firstly to generate the necessary attached biomass and then subsequently to use this as an immobilised cell system to effect substrate transformation. There has also been speculation on the possible future application of simple forms of cell immobilisation in conventional fermentations. Normal batch operation involves inefficient use of biomass because large quantities of metabolically active cells are discarded at the end of the fermentation. The use of immobilisation systems such as pellet retainment or microbial films permits reuse of cells and consequently may allow extended operation of batch processes (Abbott, 1977).

REFERENCES

ABBOTT, B.J. (1977). Immobilised cells. In *Annual Reports on Fermentation Processes*. Volume 1, pp 205-233. Edited by D. Perlman & G.T. Tsao. New York: Academic Press.

ANDERSON, G.K. & DONNELLY, T. (1977). Anaerobic digestion of high strength industrial wastewaters. *The Public Health Engineer* 5, 64-70.

ANDERSON, J.G., BLAIN, J.A., DIVERS, M. & TODD, J.R. (1980). Use of the disc fermenter to examine production of citric acid by *Aspergillus niger*. *Biotechnology Letters* (In Press).

ANTONIE, R.L. (1975). *Fixed Biological Surfaces - Wastewater Treatment*. Cleveland: CRC Press.

ARBUTHNOTT, J.P. & SMYTH, C.J. (1979). Bacterial adhesion in host/pathogen interactions in animals. In *Adhesion of Microorganisms to Surfaces* pp 165-198. Edited by D.C. Ellwood, J. Melling & P. Rutter. London: Academic Press.

ARTAVANIS, G. & TODD, J.R. (1980). Sulphite oxidation in the disc fermenter. *Biotechnology Letters* 2, 23-28.

ASH, S.G. (1979). Adhesion of microorganisms in fermentation processes. In *Adhesion of Microorganisms to Surfaces* pp 57-86. Edited by D.C. Ellwood, J. Melling & P. Rutter. London: Academic Press.

ATKINSON, B. (1973). The role of microbial films in fermentation. In *Microbial Engineering* pp 279-304. Edited by Z. Sterbacek. London: Butterworths.

ATKINSON, B. (1974). *Biochemical Reactors*. London: Pion Press.

ATKINSON, B. & DAOUD, I.S. (1976). Microbial flocs and flocculation in fermentation process engineering. In *Advances in Biochemical Engineering* Volume 4, pp 42-124. Edited by T.K. Ghose, A. Fiechter & N. Blakebrough. Berlin: Springer-Verlag.

ATKINSON, B. & FOWLER, H.W. (1974). The significance of microbial film in fermenters. In *Advances in Biochemical Engineering* Volume 3, pp 221-227. Edited by T.K. Ghose, A. Fiechter 7 N. Blakebrough. Berlin: Springer-Verlag.

ATKINSON, B. & KNIGHTS, A.J. (1975). Microbial film fermenters: their present and future applications. *Biotechnology & Bioengineering* 17, 1245-1267.

BERRY, D.R., CHMIEL, A. & AL-OBAIDI, Z. (1977). Citric acid production by *Aspergillus niger*. In *Genetics and Physiology of Aspergillus*. Edited by J.E. Smith & J.A. Pateman. London: Academic Press.

BLAIN, J.A., ANDERSON, J.G., TODD, J.R. & DIVERS, M. (1979). Cultivation of filamentous fungi in the disc fermenter. *Biotechnology Letters* 1, 269-274.

BORCHARDT, J.A. (1971). In *Biological Waste Treatment* pp 131-140. Edited by R.P. Canale. Chichester: Wiley-Interscience.

BULL, A.T. & BUSHELL, M.E. (1976). Environmental control of fungal growth. In *The Filamentous Fungi* Volume 2, pp 1-31. Edited by J.E. Smith & D.R. Berry. London: Edward Arnold.

BUNGAY, H.R., WHALEN, W.J. & SANDERS, W.M. (1969). Microprobe techniques for determining diffusivities and respiration rates in microbial slime systems. *Biotechnology & Bioengineering* 11, 765-772.

BURNS, R.G. (1979). Interaction of microorganisms, their substrates and their products with soil surfaces. In *Adhesion of Microorganisms to Surfaces* pp 109-138. Edited by D.C. Ellwood, J. Melling & P. Rutter. London: Academic Press.

CAMICI, L., SERMONTI, G. & CHAIN, E.B. (1952). Observations on *Penicillium chrysogenum* in submerged culture. 1. Mycelial growth and autolysis. Microbial growth and its inhibition. *World Health Organisation* 6, 265-275.

CHIBATA, I. & TOSA, T. (1977). Transformations of organic compounds by immobilised microbial cells. *Advances in Applied Microbiology* 22, 1-27.

CHMIEL, A. (1977). Kinetics of citric acid production by precultivated mycelium of *Aspergillus niger*. *Transactions of the British Mycological Society* 68, 403-406.

CLARK, D.S. (1962). Submerged citric acid fermentation of ferrocyanide-treated beet molasses: morphology of pellets of *Aspergillus niger*. *Canadian Journal of Microbiology* 8, 133-136.

COCKER, R. & GREENSHIELDS, R.N. (1977). Fermenter cultivation of *Aspergillus*. In *Genetics and Physiology of Aspergillus*. pp 361-390. Edited by J.E. Smith & J.A. Pateman. London: Academic Press.

COOKE, W.B. (1976). Fungi in sewage. In *Recent Advances in Aquatic Mycology* pp 389-434. Edited by E.B.G. Jones. London: Paul Elek.

CURTIS, A.S.G. (1979). Summing Up. In *Adhesion of Microorganisms to Surfaces.* pp 199-208. Edited by D.C. Ellwood, J. Melling & P. Rutter. London: Academic Press.
DURAND, G. & NAVARRO, J.M.(1978). Immobilised microbial cells. *Process Biochemistry* 13, 14-17, 20-23.
ELLWOOD, D.C., MELLING, J. & RUTTER, P. (1979). Introduction. In *Adhesion of Microorganisms to Surfaces* pp 1-4. Edited by D.C. Ellwood, J. Melling & P. Rutter. London: Academic Press.
ENGELBART, W. & ENGELBART, F. (1975). Process for making optimum chemical conversion and biological fermentations. United States Patent 3,880,716.
FINN, R.K. & FIECHTER, A. (1979). The influence of microbial physiology on reactor design. In *Microbial Technology: Current State, Future Prospects* pp 83-105. Edited by A.T. Bull, D.C. Ellwood & C. Ratledge. Cambridge: Cambridge University Press.
HUANG, M.Y & BUNGAY, H.R. (1973). Microprobe measurements of oxygen concentrations in mycelial pellets. *Biotechnology & Bioengineering* 15, 1193-1197.
KATINGER, H.W.D. (1977). New fermenter configurations. In *Biotechnology and Fungal Differentiation* pp 137-155. Edited by J. Meyrath & J.D. Bu'Lock. London: Academic Press.
KORNEGAY, B.H. & ANDREWS, J.F. (1968). Kinetics of fixed-film biological reactors. *Journal of the Water Pollution Control Federation* 40, 460-468.
KRISTIANSEN, B. (1978). Trends in fermenter design. *Chemistry and Industry* 787-789.
KRISTIANSEN, B. & BU'LOCK, J.D. The design of a fast flow reactor for aerobic treatment of dilute waste streams. *Biotechnology & Bioengineering* (In Press).
LOCKWOOD, L.B. (1975). Organic acid production. In *The Filamentous Fungi* Volume 1, pp 140-157. Edited by J. E. Smith & D.R. Berry. London: Edward Arnold.
MEADOWS, P.S. & ANDERSON, J.G. (1979). The microbiology of interfaces in the marine environment. *Progress in Industrial Microbiology* 15, 207-265.
METZ, B. & KOSSEN, N.W.F. (1977). The growth of moulds in the form of pellets - a literature review. *Biotechnology & Bioengineering* 19, 781-799.
MIURA, Y. (1976). Submerged aerobic fermentation. In *Advances in Biochemical Engineering* Volume 4, pp 3-40. Edited by T.K. Ghose, A. Fiechter & N. Blakebrough. Berlin: Springer-Verlag.
MUNSON, R.J. & BRIDGES, B.A. (1964). 'Take-over' - an unusual selection process in steady state cultures of *Escherichia coli*. *Journal of General Microbiology* 37, 411-418.

OLEM, H. & UNZ, R.F. (1977). Acid mine drainage treatment with rotating biological contactors. *Biotechnology & Bioengineering* 19, 1475-1491.

PHILLIPS, D.H. (1966). Oxygen transfer into mycelial pellets. *Biotechnology & Bioengineering* 8, 456-460.

PIRT, S.J. (1966). A theory of the mode of growth of fungi in the form of pellets in submerged culture. *Proceedings of the Royal Society B* 166, 369-373.

RUTTER, P. (1979). Accumulation of organisms on the teeth. In *Adhesion of Microorganisms to Surfaces* pp 139-164. Edited by D.C. Ellwood, J. Melling & P. Rutter. London: Academic Press.

RUTTER,P.R. & ABBOTT, A. (1978). A study of the interaction between oral streptococci and hard surfaces. *Journal of General Microbiology* 105, 219-226.

SMITH, J.E., NOWAKOWSKA-WASZCZUK, A. & ANDERSON, J.G. (1974). Organic acid production by mycelial fungi. In *Industrial Aspects of Biochemistry* pp 297-317. Edited by B. Spencer. Amsterdam: Elsevier.

TAGUCHI, H. (1971). The nature of fermentation fluids. *Advances in Biochemical Engineering* 1, 1-30.

TOMLINSON, T.G. & SNADDON, D.H.M. (1966). Biological oxidation of sewage by films of microorganisms. *Air Water Pollution* 10, 865-881.

TRINCI, A.P.J. (1970). Kinetics of the growth of mycelial pellets of *Aspergillus nidulans*. *Archives of Microbiology* 73, 353-367.

TSAO, G.T. & LEE, Y.H. (1977). Aeration. In *Annual Reports on Fermentation Processes* Volume 1, pp 115-149. Edited by D. Perlman & G.T. Tsao. New York: Academic Press.

UGOLINI, F. (1960). Fermentors a tamburo rotante di tipo modificanto. *Rendiconti Dell' Instituto Supyriore Di Sanita* 23, 819-821.

UNZ, R.F., OLEM, H. & WICHLACZ, P.L. (1979). Microbial ferrous iron oxidation in acid mine drainage. *Process Biochemistry* 14, 2-6, 28.

WHITAKER, A. & LONG, P.A. (1973). Fungal pelleting. *Process Biochemistry* 8, 27-31.

WINKLER, M.A. & THOMAS, A. (1978). Biological treatment of aqueous wastes. In *Topics in Enzyme and Fermentation Biotechnology* 2, pp 200-279. Edited by A. Wiseman. Chichester: Ellis Horwood.

YANAGITA, T. & KOGANE, F. (1963). Cytochemical and physiological differentiation of mould pellets. *Journal of General and Applied Microbiology* 9, 179-187.

YOSHIDA, T., TAGUCHI, H. & TERAMOTO, S. (1968). Studies on submerged culture of Basidiomycetes. IV. Distributions of respiration and other metabolic activities in pellets of Shiitake *(Lentinus edodes)*. *Journal of Fermentation Technology* 46, 119-124.

THE DEVELOPMENT OF IMMOBILISED FUNGAL PARTICLES AND THEIR USE IN FLUIDISED BED FERMENTERS

B. Atkinson[1] and P.J.S. Lewis[2]

[1]*Department of Chemical Engineering,*
University of Manchester Institute of Science and Technology,
Manchester.
[2]*Wander Foods Limited,*
Kings Langley, Hertfordshire.

NOMENCLATURE

C_I	inlet substrate concentration	ML^{-3}
C_o	fermenter substrate concentration	ML^{-3}
F	volumetric flowrate	L^3T^{-1}
G_{max}	maximum specific growth rate	T^{-1}
K_m	Monod coefficient	ML^{-3}
M_i	concentration of immobilised microorganisms in a fermenter	ML^{-3}
V	fermenter volume	L^3

INTRODUCTION

Continuous microbial film fermenters contain high biomass concentrations which are largely independent of throughput, do not suffer from wash-out and for most fermentations operate at low viscosities because of the segregation between the microbial and fluid phases (Atkinson & Knights, 1975). Control of the microbial film thickness is necessary to achieve a steady biomass hold-up and facilities for achieving a predetermined thickness are highly desirable. The latter requirement arises because the overall rates of substrate uptake, growth and product formation depend upon film thickness (Atkinson, 1974). In addition, it is quite possible that the overall yield coefficients also depend upon film thickness and that for many microbe-substrate systems, optimum film thicknesses exist which lead to maximum rates and/or optimum yields.

Fluidised beds of biomass support particles are an

attractive proposition for the control of film thickness by the abrasion which results from the relative movement of the particles. The use of such beds leads to possibilities for achieving any required thickness and, in addition, to good heat transfer, mass transfer and mixing characteristics.

The Completely Mixed Microbial Film Fermenter (CMMFF) was developed by Atkinson & Davies (1972) and a detailed description of an aseptic laboratory unit is given by Atkinson & Knights (1975). The basic flow diagram for the CMMFF is given in Fig. 1. The vertical tube contains a bed of immobilised biomass, e.g. inert particles covered by a microbial film, which is fluidised by the recycle pump. Feed inflow and effluent outflow take place at the reservoir at the top of the flow circuit. The reservoir also houses the control systems and pH/temperature monitoring equipment. When the recycle rate is high compared with the feed rate the total contents of the system are completely mixed. The microbial film thickness on the support particles is controlled by the abrasive action of the fluidised particles themselves. It has been demonstrated that the biomass hold-up on the surfaces of glass beads can be controlled by varying the bed expansion (Lewis, 1975).

The CMMFF is well established as a research tool and has been used extensively with mixed cultures (non-aseptic version) and yeasts (aseptic version); the basic mathematical description has also been provided (Atkinson, 1974).

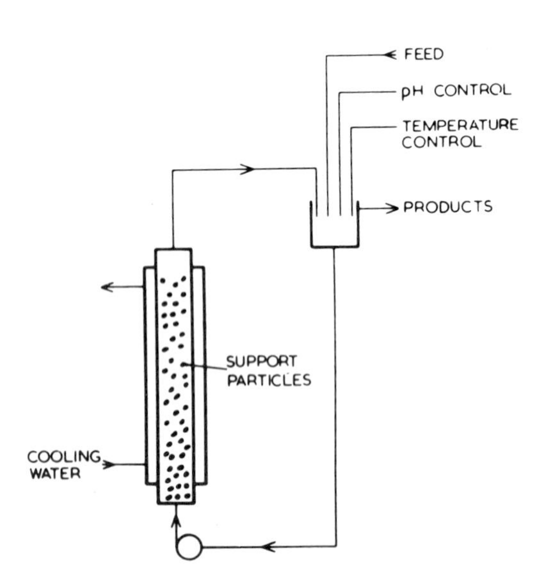

Fig. 1. The completely mixed microbial film fermenter (CMMFF).

The primary disadvantage associated with the application of microbial film fermenters to any given microbe-substrate-product reaction lies in the problems associated with the selection both of the support particles and of the physico-chemical conditions which lead to adhesion and film formation (Atkinson & Fowler, 1974).

CHOICE OF MICROBE-SUBSTRATE SYSTEM

A strain of *Aspergillus foetidus* (IMI 15954) was chosen as characteristic of fungal morphologies and because of prior knowledge (Kristiansen,1976). The ability of this organism to produce citric acid in a defined synthetic medium provides a basis for the extension of the CMMFF studies to product formation. The Shu & Johnson (1948) medium was used with glucose in place of the recommended sucrose. This substitution avoided the acid hydrolysis of sucrose to a mixture of sucrose, fructose and glucose on autoclaving, with the consequent complications of analysis and the interpretation of data; in addition citric acid yields are dependent upon the carbon source, at least when using *A. niger* (Foster, 1949). The use of glucose as a single, specifically identifiable carbon source was judged to be preferable when the purpose was the exploration of fermenter performance as opposed to maximising rates or optimising yields.

CHOICE OF SUPPORT PARTICLES

A series of shake-flask experiments were carried out for qualitative observations on fungal adhesion. These experiments involved 6 mm diameter glass spheres which were variously water, acid and caustic washed prior to prolonged incubation; no signs of adhesion were detected when the surfaces were subjected to microscopic examination. Similar results were obtained when using spheres with very rough surfaces, i.e. sand-blasted glass spheres and polyester beads with an inert silica filling. These results were compatible with those of Sanmugasunderam (1975), who found that fungal organisms formed an unsatisfactory film in spite of subjecting glass beads to a wide variety of different treatments.

Sanmugasunderam (1975) had some limited success with the adhesion of *A. niger* mycelia onto knurled stainless steel. However, since solid metal spheres have too high a density for efficient fluidisation, relatively light spheres were fabricated by compressing aluminium foil in a rivet punch and the shake flask experiments were repeated with the 'new' particles after they had been degreased in an acid wash. In these experiments the mycelia was found to grow most satisfactorily in the crevices of the compressed foil, where protection from shear and abrasion existed, while growth beyond the particle was removed. These observations lead to the concept of a particle of compressed woven stainless steel mesh and in experiments with such particles, the mycelia was found to grow within the mesh while any overgrowth was abraded from the external surfaces (B.P.A. 1978; Atkinson *et al.*, 1978).

This development changed the geometric arrangement from one based on microbial films of unknown thickness to one based upon microbial flocs of uniform, predetermined size. The latter facility arises because the floc size can be altered by merely changing the overall size of the mesh particle. It follows that a Completely Mixed Microbial Floc Fermenter version of the CMMFF was necessary.

DEVELOPMENT OF THE MESH PARTICLE - CMMFF SYSTEM

The solution found to the support surface problem involved a mesh particle in which mycelia can grow in a protected environment. For research applications stainless steel particles are to be preferred because they can be prepared easily in a variety of shapes, sizes and bulk densities, and the internal structure can be controlled by the extent to which they are compressed. Such particles have the advantage that they can be sterilised thermally, while cost is their major disadvantage.

Shake flask experiments were conducted with 3,6 and 9 mm particles to establish any size limitations which might exist in this range. Dissection of the particles after 5 days incubation showed all the particles to be completely full of mycelia. This result contrasted with similar experiments by Perez (1977) who used layers of stainless steel mesh wound to the outside of a stainless steel draught tube in a continuous air-lift fermenter. On termination of his experiments Perez (1977) found no growth on the inner layers. One difference between the mesh particles and layers of mesh on a support surface lies in the fact that with the former growth can penetrate from all sides, leading to an essentially homogeneous monolithic mass of mycelia without planes of weakness such as that presented by a support surface.

The CMMFF (Atkinson & Knights, 1975; Lewis, 1975), had previously been developed for use with 6 mm glass particles, and the recycle pump capacity for fluidisation and the distributor plates had been chosen accordingly. Thus, for the present application 6 mm diameter mesh particles of 80% porosity were chosen for reasons of compatibility, leading to a similar particle density and size to the glass spheres. The preliminary shake-flask experimentation was carried out using particles prepared in a hand press developed for the purpose; however, some thousands of particles were required for the CMMFF and these were manufactured by Knitmesh Limited by compression of weft-knitted stainless steel 'stocking', with the result that each particle consisted of a single strand of wire (Fig. 7).

Equipment

The fermenter (Fig. 2) consisted primarily of a gas disengagement reservoir, pump and bed of particles. In the main reservoir were the probes for a pH meter, a contact thermometer and a heater; serum caps allowed addition of antifoam and inoculation. The pump was a stainless steel positive displacement gear pump and the recirculation rate and hence bed expansion was controlled using by-pass valves. The flow circuit was from the reservoir to the pump, then via a rotameter into the bottom of the bed and then back to the reservoir. Aseptic sample points were provided below and above the bed.

When the fermenter was operated continuously the medium was made up as a concentrate which was metered into a feed-mix vessel along with sterile distilled water. After mixing by magnetic stirrer the feed was forced by differential pressure through feed rotameters into the inner chamber of the main reservoir. The outflow from the fermenter was by overflow from the inner to the outer chamber of the main reservoir and on to an outflow steriliser which prevented back growth and contamination.

Sterile water for the feed was obtained by condensing steam. Steam from the main was first separated in a stainless still cyclone separator, then condensed and passed via a steam trap into the bottom of a vertical sterile water reservoir. Water for the feed was drawn off half-way up the column, cooled and filtered in a gamma 12 filter to remove any colloidal iron present. The water was then metered into the feed-mix vessel. Overflow of hot sterile water ran by gravity to the outer vessel of the main reservoir, and caused a large flow, which kept the drain line from becoming blocked by the growth of organisms.

Air was provided under pressure from a main, or from bottled supplies. After pressure regulation and crude filtering, the air was sterilised using ceramic candle filters. The air flow rate was controlled by a needle valve, measured on a rotameter and passed through a humidifier fed with sterile water before being injected into the circulating liquid just after the rotameter and beneath the bed. The outlet air was passed through a steriliser (to prevent entry of contaminants

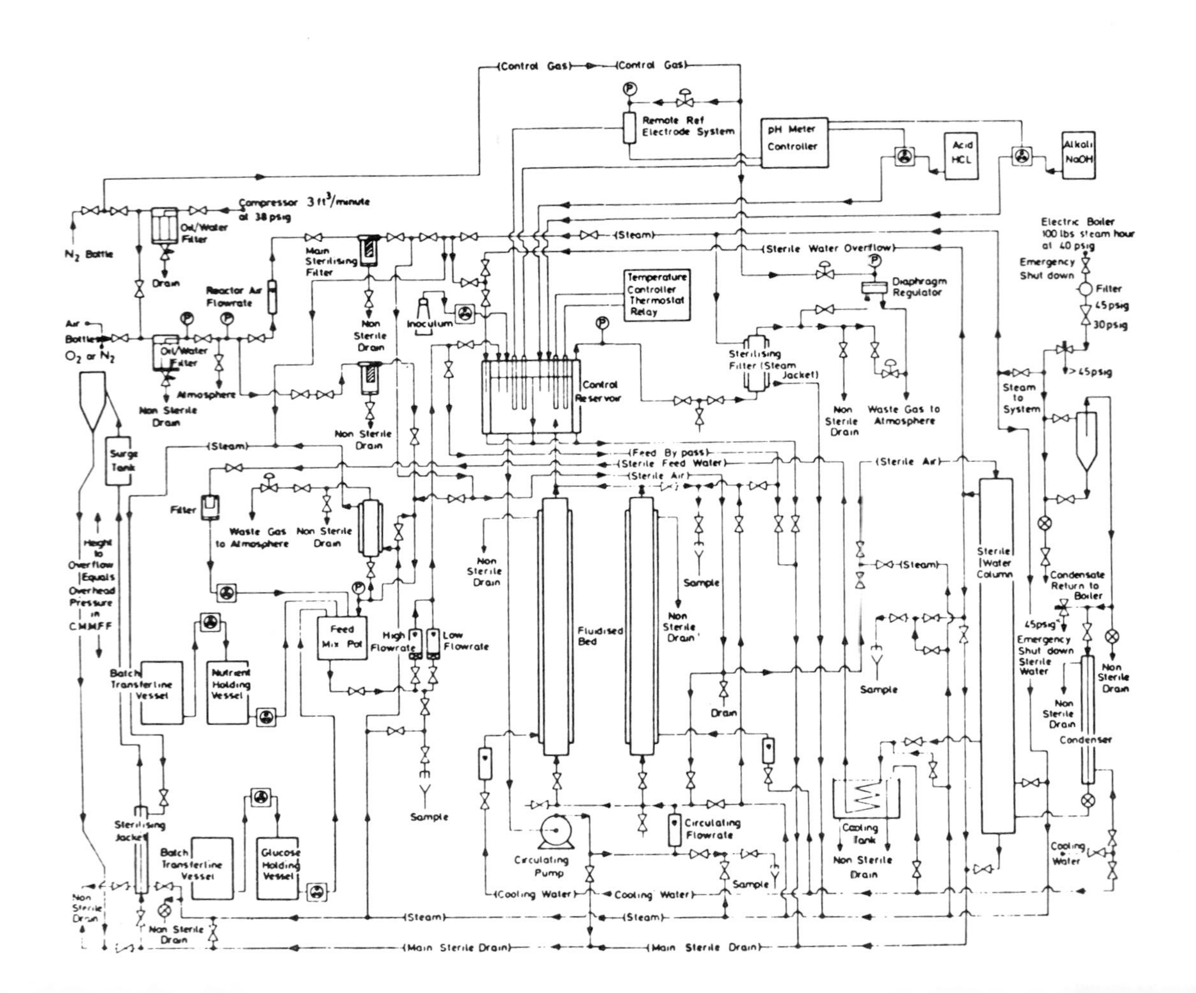
(Control Gas)
(Control Gas)
Remote Ref Electrode System
pH Meter Controller
Acid HCL
Alkali NaOH
Compressor 3 ft^3/minute at 38 psig
Oil/Water Filter
N_2 Bottle
Drain
Main Sterilising Filter
(Steam)
(Sterile Water Overflow)
Electric Boiler 100 lbs steam hour at 40 psig
Emergency Shut down
Filter
45 psig
30 psig
> 45 psig
Reactor Air Flowrate
Air Bottles O_2 or N_2
Atmosphere
Non Sterile Drain
Inoculum
Temperature Controller Thermostat Relay
Diaphragm Regulator
Sterilising Filter (Steam Jacket)
Waste Gas to Atmosphere
Control Reservoir
Steam to System
Surge Tank
Filter
Waste Gas to Atmosphere
(Feed By pass)
(Sterile Feed Water)
(Sterile Air)
Sample
Sterile Water Column
Condensate Return to Boiler
Height to Overflow Equals Overhead Pressure in C.M.M.F.F.
Feed Mix Pot
High Flowrate
Low Flowrate
Fluidised Bed
45 psig
Emergency Shut down Sterile Water
Batch Transferline Vessel
Nutrient Holding Vessel
Condenser
Sterilising Jacket
Glucose Holding Vessel
Circulating Pump
Circulating Flowrate
Cooling Tank
Cooling Water
(Cooling Water)
(Main Sterile Drain)

by this route) to a back pressure regulator. The fermenter was held under pressure to reduce the possibility of entry of contaminants.

The circulating liquid was cooled by a jacket around the bed which slightly overcooled the system and a thermostat/heater in the main reservoir controlled the temperature.

As the apparatus was too large to be dismantled and autoclaved, steam at 25 p.s.i.g. was piped to various parts of the system. The whole apparatus was steam sterilised at 15 p.s.i.g. for 2 to 3 h prior to inoculation, steam was used at all the sample points, inlets and outlets to prevent the ingress of contaminants.

Commissioning

Preliminary experiments using glass spheres The CMMFF containing 9000 x 6 mm diameter glass spheres was subjected to a number of continuous fermentation experiments during which it was concluded that, when using fungal organisms, neither upper nor lower distributor plates could be used because relatively large flocs caused blockages. These flocs were usually generated by the occasional break-away of film growth from the ancillary lines and vessels of the fermenter. The blockage problem was overcome by removing the distributor plates and the circulating fluid rotameter, and by constructing the recirculation loop of 0.5 inch diameter stainless steel line to reduce adhesion and provide an adequate scouring velocity. The particles were retained in the bed by the construction between the recirculation line and the 50 mm diameter column. Infrequent blockages still occurred and to maintain operation on these occasions a mechanical disrupter was installed at the base of the bed. The disrupter consisted of a metal bar which entered the column through a sterile gland and which could be moved, as required, back and forth along the column axis within the bed of particles.

Preliminary experiments using empty mesh particles The 9000 empty mesh particles and initial fluidisation trials were carried out using distilled water. The commercially fabricated particles were found to be free of entanglement, one with another, compared with the hand produced woven mesh particles, even so they were more difficult to fluidise than the glass spheres. Due to the corregated nature of the surfaces relative motion between the mesh particles only took place after the particles separated and fluidisation was very unstable, with particles in the centre of the bed moving vigorously and

Fig. 2. (Opposite) Flow diagram of the aseptic CMMFF system.

erratically while those adjacent to the column walls were less mobile. The particles at the base of the bed were stationary suggesting compression by the weight of the bed.

To alleviate compression the number of particles was reduced to 5250 and these were easily fluidised. It was found that greater bed expansion occurred when the recirculating liquid was replaced by growth medium, a phenomenon noticed on previous occasions when using glass spheres.

Preliminary experiments using biomass filled particles The CMMFF was first operated batchwise until growth was apparent and then continuously at an inlet concentration of 80 gl^{-1} glucose. After two days of continuous operation fluidisation virtually ceased. At this stage the particles were essentially full of biomass and while there was some tendency for packs of particles to form, there was no tendency for the particles to grow together. It was further observed that the fuller the particles the higher they collected in the bed with the empty particles at the bottom (Fig. 3). This 'classification' is perhaps the reverse of that expected on purely density considerations and suggests that the fluid passes through the particles which are less than full.

Dismantling the equipment revealed that particles had been carried into the narrow bore manifolding above the column causing constriction and a reduced recirculation flow; some 20 particles had been carried 70 cm above the top of the bed. Observations under operating conditions had revealed that occasionally an empty particle would become entrained by an air bubble, while filled particles were unaffected. Subsequently extra care was exercised when aerating beds of partially filled particles.

In view of these various problems the 50 mm internal diameter column was replaced by one of 75 mm internal diameter to reduce the 'packing' effect, while the number of particles was reduced to 3800 to increase the distance between the top of the bed and the manifolding. The initial experiment with the modified CMMFF was carried out in the absence of adequate cooling facilities since they were, at the time, unavailable; this resulted in temperatures of 40°C. During this experiment all the particles moved freely and no packing occurred but a number of large buoyant particles were observed with mycelia growing beyond the mesh boundaries. Overgrown particles were not seen at any time with the 50 mm column when the temperature was maintained at 25°C. Once overgrowth had taken place it could not be overcome by the violent agitation produced by increasing the air flow rate. Subsequent experimentation at controlled lower temperatures using the 75 mm column led to the particles being filled with biomass but without significant overgrowth.

Fig. 3. Biomass filled mesh particles in the CMMFF.

General Design Summary

(a) Distributor plates cannot be located at the top and bottom of the bed as breakaway flocs impact on them and restrict the flow. Sedimentation/fluidisation principles must be relied on to maintain the particles in the column. Further the system should contain as few right angle bends and constrictions as possible to reduce the possibility of blockage.

(b) For laboratory scale experimentation it is useful to have a mechanical disruptor at the base of the bed, in case breakaway flocs cause blockage, though blockage is unlikely if fluidisation is maintained at all times.

(c) Empty mesh particles have fluidisation characteristics which make them difficult to suspend and careful control is required.

(d) It is impractical to have very tall beds when using stainless steel mesh particles, as the particles at the bottom become slightly compressed and interlock, rendering fluidisation difficult; shorter, wider beds are to be preferred.

(e) Empty mesh particles tend to entrap air, become buoyant and can be washed from the column. Care is required in controlling the aeration rate when empty particles are present.

(f) A length of empty column above the bed allows entrapped air to escape from the particles and causes them to fall back into the bed.

(g) The ratio of bed diameter to particle diameter should be at least 12:1. Smaller ratios cause aggregations of particles and a tendency to form packs which may be pushed up the column rather like a plug. The wider the bed for a given particle size, the less is the likelihood of packs developing and if they do form the greater is the possibility of break-up before they achieve an unacceptable size.

(h) A high fluid velocity through the manifolding helps to prevent wall growth and subsequent blockage.

(i) Optimum bed expansion for attrition of particles appears to be in the range 10-15%, beyond this the greater separation of the particles appears to result in less attrition and greater overgrowth.

FLUIDISATION MECHANICS

The initial development of the CMMFF system was carried out using glass spheres and the changeover to mesh particles produced more problems than originally envisaged. Fluidisation of the whole bed was difficult. In particular, a packing effect often occurred at the top of the bed and once formed tended to extend downwards. Increased recirculation failed to cure this phenomenon and the need for preliminary study of the fluidisation mechanics of the mesh particles became apparent.

A small test rig was set up as shown in Fig. 4 to study particle behaviour. This apparatus enabled experiments on the relationship between pressure drop, bed expansion and flow-rate to be carried out quickly and easily. Flowrate/pressure drop data was obtained with an empty bed, 500, 1000 and 1500 glass particles, then repeated with both empty and biomass filled mesh particles.

The pressure drop/flow rate data are plotted in Fig. 5. As expected, the curves for the glass spheres show classical fluidisation characteristics, i.e. a sharp rise in pressure drop as the flowrate through the static bed is increased and a maximum pressure drop as the bed starts to fluidise, this is followed by a slight drop once the particles have

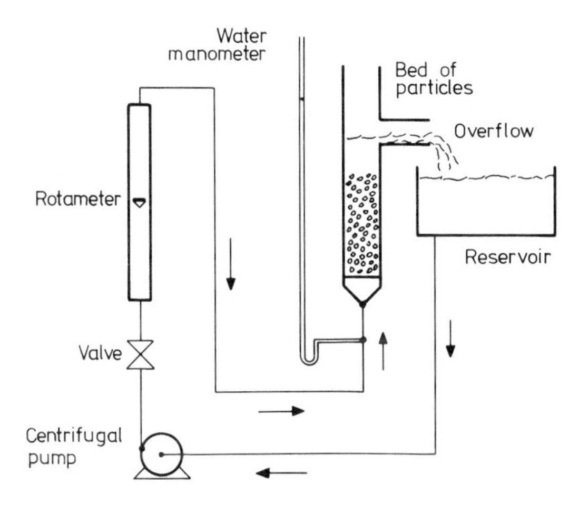

Fig. 4. Rig for fluidisation tests.

started to move, leading to a constant pressure drop independent of flowrate and equal to the weight of the bed (Coulson & Richardson, 1968). However, there is little similarity with the classical data beyond the point of incipient fluidisation for the 1000 and 1500 empty mesh particles; the maximum is greatly exaggerated and there is no evidence of a constant pressure drop over the range of flowrates studied. The experiment with 500 empty mesh particles appears to be anomalous, possibly due to experimental error. When the particles are filled with biomass the shape of the curves is as expected. It is of interest to note that the data for empty and filled particles is similar below fluidisation and that for glass spheres and filled particles beyond fluidisation, this indicates that surface effects predominate below fluidisation and that the internal structure exerts a significant influence when the particles are in the fluidised state.

Clearly the differences in fluidisation properties call for careful operation during start-up, i.e. the period when the particles are accumulating biomass.

In an attempt to understand the packing effect which occurred with relatively tall beds of mesh particles, an experiment was carried out in which the number of particles in the bed, and hence the bed height, was progressively increased at a constant flow rate beyond the incipient fluidisation velocity. The pressure drop/number of particles data is given in Fig. 6. The slope of each of the curves is

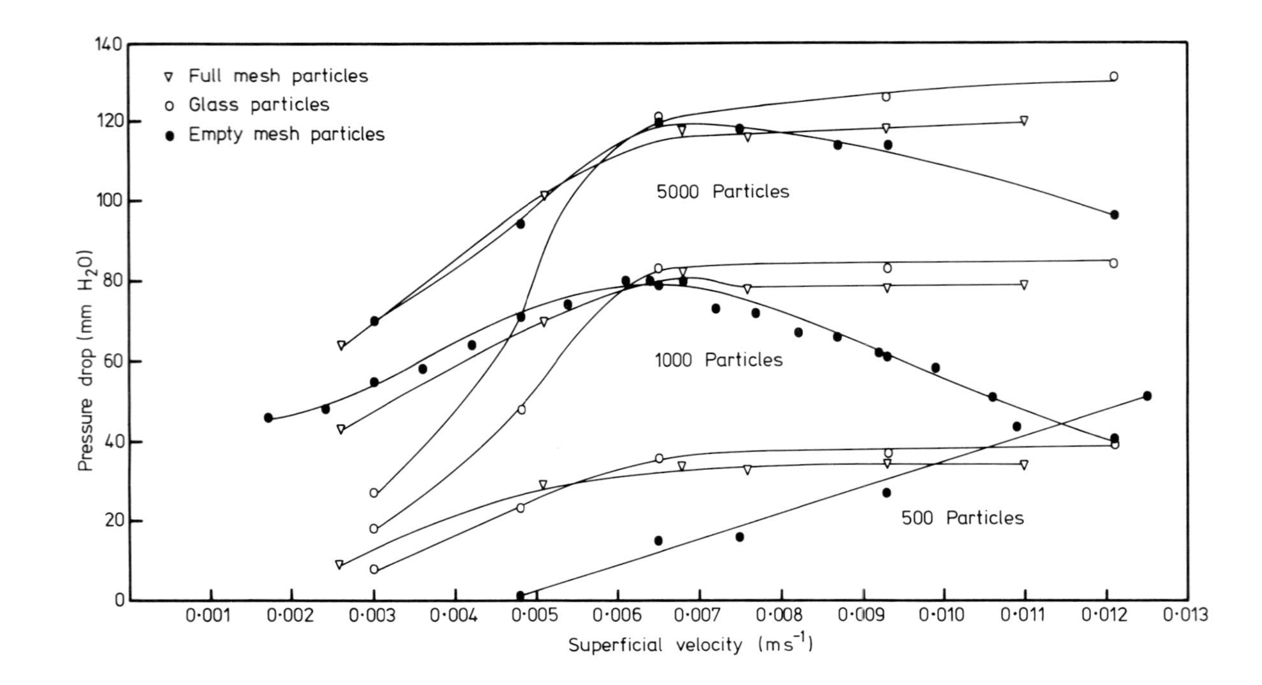

Fig. 5. Pressure drop versus superficial liquid velocity (50 mm diameter bed).

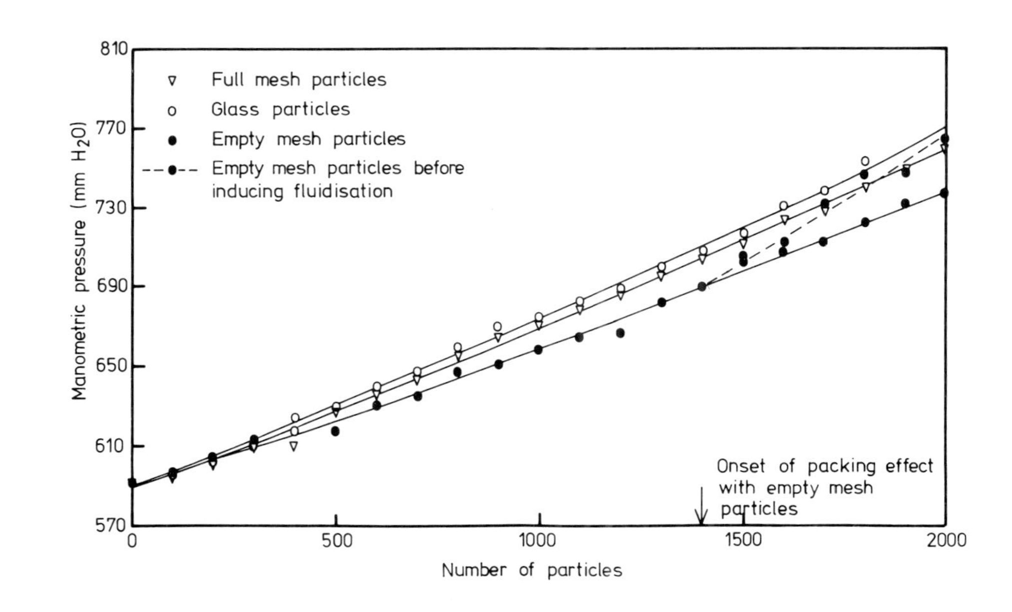

Fig. 6. Pressure drop versus number of particles (50 mm diameter bed flowrate of 11 l/min and superficial velocity 0.0093 ms^{-1})

essentially constant and for the empty mesh particles is less than that for the glass spheres and full particles as expected from Fig. 5. However, in the region of 1500 empty mesh particles an anomally occurred, fluidisation partially ceased due to the 'packing' effect and the slope of the pressure drop data increased. When full fluidisation was induced, by temporarily increasing the flowrate on each occasion that the 'packing' effect occurred, the pressure drop/number of particles relationship was compatible with that for small numbers of particles. It was also noted that the length of the 'packed' zone was greater the larger the number of particles in the bed. Although the mesh particles consist of a single strand of stainless steel wire and the two ends have little chance of becoming entangled it can be envisaged that adjacent particles become temporarily enmeshed due to the corregated nature of their surfaces.

The simplest solution to the packing problem is to reduce the height of the bed. Another possible solution is to add smooth particles to the bed to reduce interlocking. A single experiment was performed using equal numbers of glass spheres and empty mesh particles. At fluid velocities beyond that of incipient fluidisation there was no tendency to classify, but in the region of incipient fluidisation the heavier glass particles moved to the bottom of the bed. When classification had taken place remixing could not be achieved even at high fluid velocities.

Classification also occurred with mixtures of glass sphered and biomass filled particles, where the densities are very similar since the average weight of a biomass filled particle is 0.30 g while that of a glass sphere is 0.29 g. Clearly while smooth particles could be developed on a size and density basis which would not classify under any conditions when mixed with empty mesh particles the margin for error is small and as such makes this possibility impracticable.

PARTICLE CHARACTERISTICS

Morphology

In the shake-flask experiments where all the particles were exposed to the same conditions of shear, abrasion etc. the particles had the appearance of that shown in Fig. 7, i.e. the supporting mesh was visible and full of mycelia. The great majority of the particles in the fluidised bed also had the appearance of that in Fig. 7a. To confirm that the particles were full of biomass one of them was bisected (Fig. 7b), and another unravelled. Another less direct, but more quantitative method was also adopted with the same

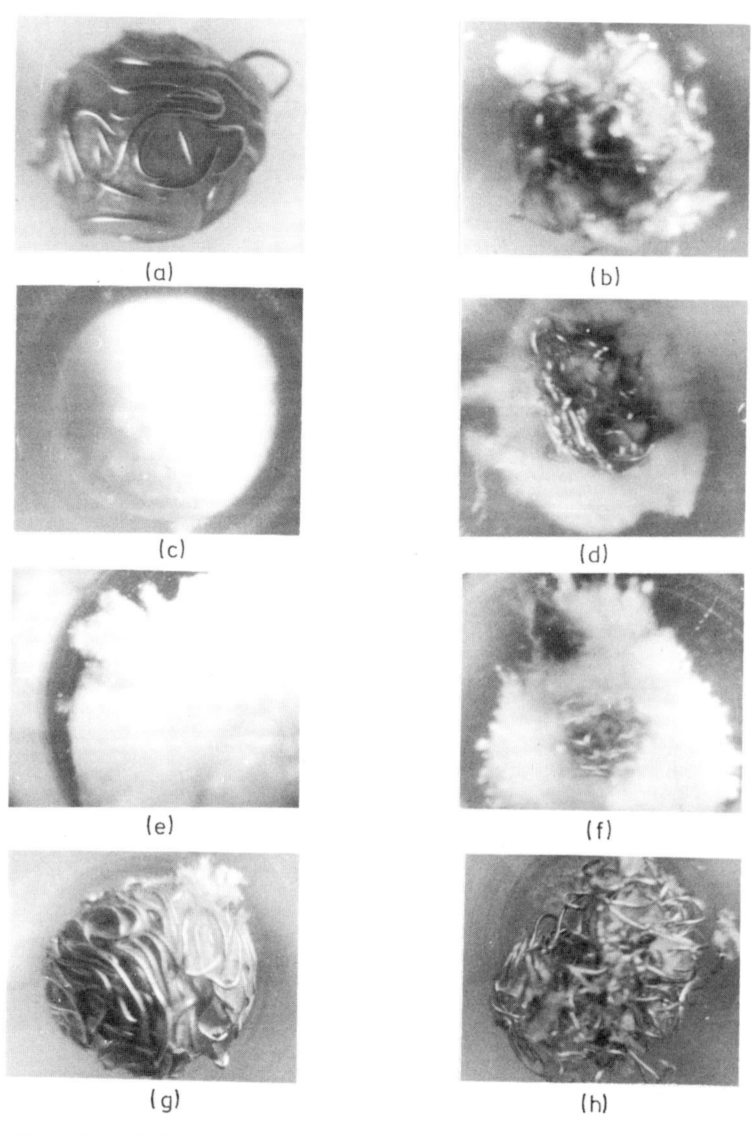

Fig. 7. Particle morphologies in the CMMFF

(a) Typical morphology (c) 'Orange-peel' morphology
(e) Fluffy morphology (g) Dried particle
(b)(d)(f)(h) Bisected particles corresponding to (a)(c)(e)(g) respectively.

objective. Since the weight and porosity of the empty and full particles was known, as was the density of wet and dry biomass, it was possible to calculate both the actual and potential biomass hold-ups within a particle (see Section *Biomass Hold-Up*).

Morphologies other than that in Fig. 7a were present under certain conditions. Upon start-up, when all the particles were empty and fluidisation presented difficulties, growth always started at the top of the bed and progressed downwards, probably due to bed porosity effects. Hence, before a steady biomass hold-up was reached, there was a distribution of particles with full ones at the top and empty ones at the bottom. There was no tendency to mixing because empty and full particles have different fluidisation characteristics leading to segregation. Upon prolonged operation overgrown particles (Fig. 7c) appeared at the top end of the bed. These particles were spherical, discrete and overgrown and had an 'orange-peel' morphology as may be seen from the bisection in Fig. 7d. Particles of this type occurred in the top 2-3 cm of the bed presumably due to reduced attrition and the enhanced buoyancy consistent with reduced density. When the bed expansion exceeded 10-15% the overgrown particles appeared lower down the bed and at very high expansions the whole bed consisted of particles with the 'orange-peel' morphology.

When the temperature control was ineffective, and the temperature rose to 35-40°C, gross overgrowth of the particles occurred (Fig. 7e). Fig. 7f is a bisection of this type of particle and shows the relative size of the mesh particle to the overgrowth. This overgrowth was unlike the 'orange-peel' morphology, being very loose and fluffy, totally collapsing when the particle was removed from water. These particles appear to be an extension of the 'orange-peel' condition at high temperatures. The large fluffy particles were quite buoyant and on occasion floated out of the column (over 70 cm free column space) into the narrower manifolding causing blockage. Gas entrapment did not appear to be the cause of the buoyancy since particles with gas included rose quickly, whereas the overgrown particles floated gently upwards (and sometimes downwards) in the free column space.

Biomass Hold-Up

The 80% particle porosity was achieved by inserting a known weight of mesh of known density into a die of known volume. The average weight of an empty particle was 0.1689 g while the average weight of a wet, washed and drained particle full of *A. foetidus* mycelia was 0.3007 g leading to a wet biomass hold-up of 0.1318 g. On drying full particles the average weight was found to be 0.1733 g leading to a dry biomass hold-up of 0.0044 g.

For 6 mm particles of 80% porosity the free volume is 0.0905 cm^3 and since the wet biomass density was found to be 1.0019 cm^{-3} the corresponding potential wet biomass weight for a particle is 0.0965 g.

The measure of agreement between the actual and potential wet biomass hold-ups confirms that the particles were essentially full.

Cleaning

It is clearly necessary to have available a procedure for emptying particles of biomass following experimentation. The first attempt to achieve 'clean' particles was to leave them in the fermenter and to fluidise vigorously using distilled water, but even after 7 days no significant loss of mycelium was evident. Addition of HCl (to pH 1) or NaOH (to pH 10) via the pH control system, while continuing fluidisation, also failed to lead to loss of mycelium. Placing particles in a bunsen flame caused severe damage as well as discolouration of the particles without destroying the biomass, while drying at 20°C, 60°C and 105°C caused hard, dry mycelia pellets to form but with little shrinkage or tendency to detach (Figs. 7g,h). Wetting dried particles reconstituted the biomass without other effect. Placing particles in a baffle furnace at 500°C reduced the biomass to ash which was difficult to remove from the bulk of the particle. Immersion in concentrated acids, alkalis and disinfectants, such as lysol, were tried for periods of up to two weeks but without success. Finally, strong solutions of hypochlorite bleach were found to clean some 90% of the particles in 6 days.

PERFORMANCE OF THE MESH PARTICLE - CMMFF SYSTEM

The CMMFF was sterilised, filled with media and inoculated with spores pre-germinated in a shake-flask, the feed was then applied and the system sampled until a steady state was achieved. The dilution rate was then set at a new value, a new steady state achieved (usually after 2 to 3 days), and so on. The inlet glucose concentration was held as near constant as possible over a range of dilution rates. Experimental data on the fractional conversion (C_o/C_I) are plotted versus dilution rate (F/V) in Fig. 8 for various values of the inlet substrate concentration (C_I).

The fractional conversion data in Fig. 8 shows a different response to dilution rate from that associated with the chemostat, in particular there is no evidence of 'wash-out' in the

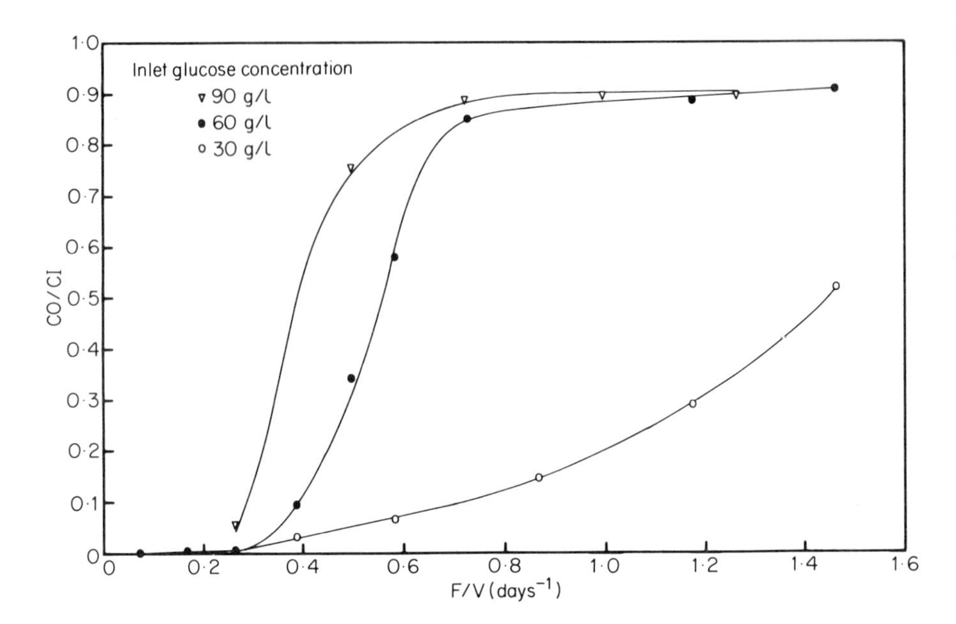

Fig. 8. Conversion efficiency versus dilution rate for the CMMFF containing 3800 x 6 mm mesh particles

conventional sense and there is a dependency on the inlet concentration. The theory for the CMMFF developed by Atkinson & Davies (1972) for particle supported microbial films in the absence of 'solid' or liquid phase diffusion limitations suggests that

$$\frac{C_o}{C_I} = g\left(\begin{matrix}\text{dilution,} & \text{number of,} & \text{inlet substrate} \\ \text{rate} & \text{particles} & \text{concentration}\end{matrix}\right)$$

Fig. 9 is a typical example of the performance characteristics of a CMMFF containing a given amount of immobilised biomass in the form of films and when operated at various inlet substrate concentrations over a range of dilution rates. In qualitative terms the agreement between Figs. 8 and 9 is excellent and would not be changed in principle either by the use of mesh particles or by their being substrate diffusion limited.

The numbers of particles in the system could not be changed as previously when using yeast on glass support particles, *i.e.* by using four columns and a bypass with each of the

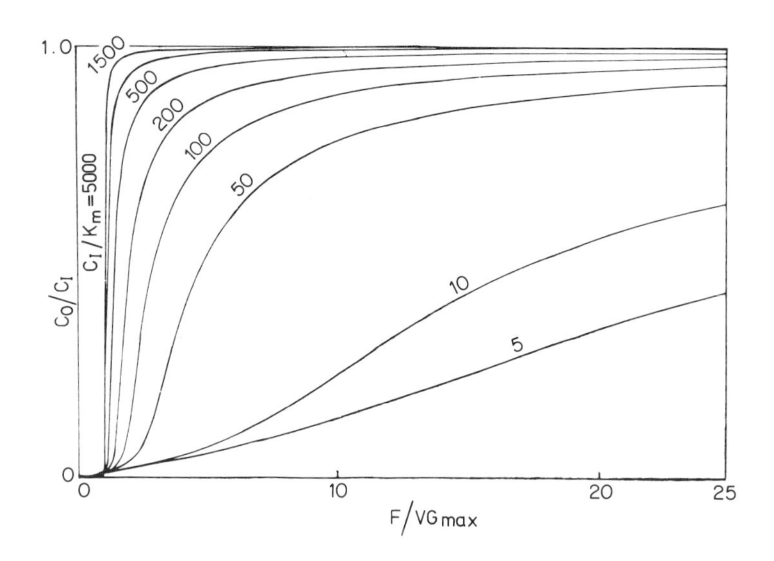

Fig. 9. Performance of a completely mixed fermenter

columns isolated in turn at its base (Lewis, 1975), because of the sensitivity of the system to blockage.

Although more difficult to carry out a particle removal procedure had to be adopted. This method involved isolating the column top and bottom, and with flaming removing the disruptor gland, and then allowing particles to fall out. With further flaming the gland was replaced and the system sealed, recirculation restored and the experiment continued. This method enabled data points to be obtained, by progressive and repeated removal of particles. Considerable risks of contamination were involved, but after three attempts where contaminants were detected, the technique was perfected, and the data recorded in Fig. 10 were obtained. These data correspond to a high flowrate, *i.e.* a region of low conversion, and show the improvement in the conversion efficiency with number of particles in the fermenter. The linear relationship suggests that the overall rate of substrate uptake was independent of glucose concentration, *i.e.* zero order kinetics prevailed.

CONCLUDING REMARKS

The Completely Mixed Microbial Film Fermenter of Atkinson & Davies (1972) has been developed into a Completely Mixed Microbial Floc Fermenter. The modification to the original

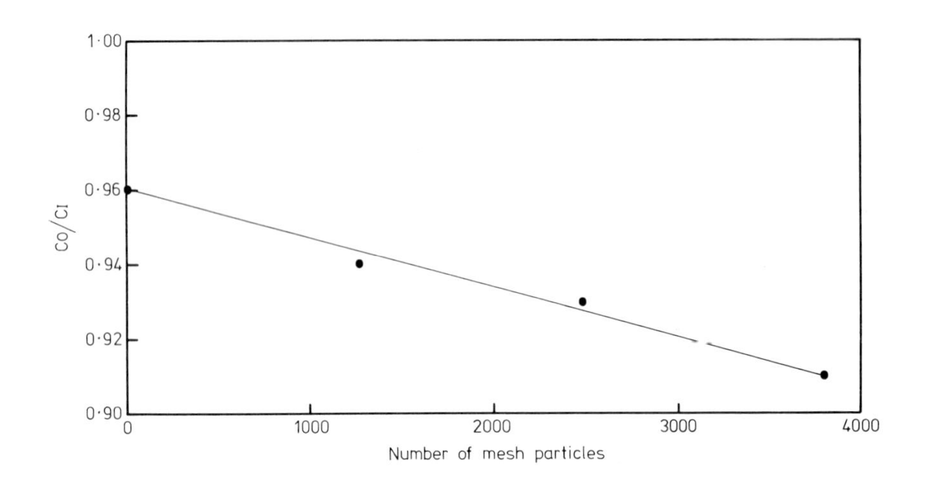

Fig. 10. Conversion efficiency versus number of mesh particles for the CMMFF (inlet glucose concentration 60 gl^{-1}, F/V 1.46 day^{-1})

concept brings all the potentially advantageous properties of the CMMFF as identified by Atkinson & Knights (1975), that much closer to realisation within biological process engineering practice. Among these advantages are high biomass hold-ups which are predetermined and independent of flow rate, absence of wash-out, segregation of biomass and liquid phases, potential for biomass recovery at low water contents, as well as good heat and mass transfer. In addition the possibility arises for utilising the diffusion limitations which occur within biomass to achieve maximum rates and optimum yields.

The development, experience and direct experimental results obtained with the CMMFF up to the present are largely supportive of the process engineering objectives which have been identified. The mesh particles advance greatly the original propositions which relied on knowledge of adhesion to produce microbial films and fluidisation phenomena to control the film thickness. Further extensive experimentation would have been necessary to provide the design engineer with the necessary correlations and enabling technology to produce the film thickness necessary to meet a given process engineering objective. Clearly any characteristic size of biomass can be produced by using mesh particles of appropriate diameter. It can reasonably be argued that low overall yield coefficients can be achieved with particles exposed to low substrate which can result from diffusion limited biological reactions; others occur in the areas of substrate inhibited systems, optimal internal environmental conditions, and product formation etc.

REFERENCES

ATKINSON, B. (1974). *Biochemical Reactors*. Pion Limited, London.

ATKINSON, B., BLACK, G.M., LEWIS, P.J.S. & PINCHES, A. (1979). Biological particles of given size, shape and density for use in biological reactors. *Biotechnology and Bioengineering*. 21, 193-200.

ATKINSON, B. & DAVIES, I.J. (1972). The Completely Mixed Microbial Film Fermenter. *Transactions of the Institute of Chemical Engineers* 50, 208.

ATKINSON, B. & FOWLER, H.W. (1974). The significance of microbial film in fermenters. *Advances in Biochemical Engineering* 3 221.

ATKINSON, B. & KNIGHTS, A.J. (1975). Microbial film fermenters - their present and future application. *Biotechnology and Bioengineering* 17, 1245.

BRITISH PATENT APPLICATION Number 38267 and similar foreign applications.

COULSON, J.M. & RICHARDSON, J.F. (1968). *Chemical Engineering* Volume 2, Pergamon Press.

FOSTER, J.W. (1949). *The Fungi*, Academic Press.

KRISTIANSEN, B. (1976). Production of organic acid by continuous culture of fungi. *PhD Thesis* University of Manchester.

LEWIS, P.J.S. (1975). Microbial film thickness in a CMMFF. *MSc Thesis* University of Wales.

PEREZ, F. (1977). Measurement of the ideal microbial film thickness for product formation. *MSc Dissertation* University of Manchester.

SANMUGASUNDERAM, V. (1975). The production of citric acid by a submerged microbial film. *MSc Thesis* University of Wales.

SHU, P. & JOHNSON, M.J. (1948). Citric acid production by submerged fermentation with *Aspergillus niger*. *Industrial Engineering Chemistry* 40, 1202.

SOLID STATE FERMENTATION AND THE CULTIVATION OF EDIBLE FUNGI

W.A. Hayes

*Department of Biological Sciences,
The University of Aston in Birmingham,
Birmingham B4 7ET*

INTRODUCTION

It is only in comparatively recent times that the principles underlying the artificial methods of mushroom culture have been fully understood and with it an appreciation of the potential of edible fungi as a food which can more usefully complement staple foods. Most of our knowledge is based on the methods which apply to the artificial culture of *Agaricus bisporus*, the common edible mushroom, but techniques of cultivating other edible species are now being perfected and are thus diversifying the opportunities of exploiting a wide range of agro-industrial wastes as substrates.

The techniques for the artificial cultivation of *Agaricus bisporus* were discovered in France in the seventeenth century when it was found that a crop of mushrooms usually resulted following the watering of melon beds made from horse manure, with spore-laden water used for cleaning mushrooms harvested from the wild. Also, it was observed that for mushrooms to form it was necessary for the beds to be "capped" with a layer of soil. In modern techniques of culture, these seemingly essential requirements for the completion of the *Agaricus bisporus* life cycle are provided as a compost, which supports the vegetative or mycelial stage of growth and casing soil, in which the transition from the micro-mycelial form to the macro-reproductive form occurs. Other microorganisms are intimately associated with growth and their activity in both substrates play a crucial part in the productivity of cultures. It is this requirement for substrates, with contrasting physical, chemical and biological properties, which distinguishes the solid-state fermentation in mushroom culture from the more common examples, e.g. silage making.

Most of the edible species of mushrooms are Basidiomycetes with low growth rates and relative to other fungi are poor competitors. The methods employed in artificial culture are therefore designed to give the fungus a competitive advantage over other organisms that may be present. Aseptic techniques are, however, applied to the first stages of the process, namely the production of inoculum or spawn.

There are a wide range of edible species which occur naturally, some of which, such as the Truffles and Morels, are greatly appreciated in sophisticated cuisine, but only a few are at present amenable to the artificial techniques of culture. A smaller number have assumed economic importance, and these serve as good illustrations of the range of substrates that may be exploited, the physiological capacities of edible species and their capacities to colonise naturally occurring materials at different stages of decomposition. These may be grouped as follows:

Group I typified by *Agaricus bisporus*, the common cultivated mushroom, for which plant residues (usually straw) are enriched to enhance substrate decomposition by composting.

Group II typified by *Volvariella volvacea*, known as the padi straw or Chinese mushroom, which grows on plant residues, usually rice straw, which does not require enrichment or extensive decomposition by composting.

Group III typified by *Lentinus edodes* or the Japanese Shiitake which grows on freshly felled wood of several tree species.

CULTURE OF *AGARICUS BISPORUS*

Ecology of mushroom culture

A meaningful understanding of mushroom culture - a batch fermentation, may best be achieved by following the microbial composition of substrates at different stages in the process. From this ecological viewpoint the stages are clearly defined:

Compost preparation
Inoculation and incubation
Induction of fruitbody formation
Cropping and its termination.

Compost preparation Historically, partially decomposed horse manure has been the principal medium for providing the required nutrients in the artificial culture of *Agaricus* mushrooms and it is only in comparatively recent times that other materials have been used successfully. It is unlikely, however, that the exploitation of other materials would have

been possible without the knowledge and experience gained in the preparation and composting of horse manure.

Most of the improvements in methods of compost preparation have stemmed from the empirical trials and experiences of commercial cultivators, but during this century important contributions have been made following the studies of Waksman and his co-workers in the 1930's, who related the changes in the composition during composting to the activity of micro-organisms and provided the basis for considering its preparation as a fermentation involving a mixed culture of micro-organisms.

The enrichment of horse manure with substances rich in protein, e.g. cotton seed meal and chicken manure, has been common practice for many decades, but studies by Lambert & Davis (1934) identified four zones of activity in composting manure which were related to the natural diffusion of air into the stacks and to microbial activity. In a later study Lambert (1941) demonstrated that compost produced under aerobic conditions at temperatures between 50-60°C was the most suitable for mushroom growing, which suggested a direct relationship with the activity of specific groups of microorganisms. This led to the establishment of what is known as peak-heating of compost, which is essentially a pasteurisation treatment during which temperatures are held within this range.

Pizer (1931) and Pizer & Thompson (1938) found that the addition of calcium sulphate to compost caused the colloids to flocculate which resulted in a more granular structure with increased water-holding capacity and which resulted in improved mushroom growth at a later stage of the process.

In considering the practical aspects of composting, the introduction of mechanical aids to replace hand turning has also contributed much to improvements; relatively rapid turning of the compost stacks ensures more uniform aeration, more adequate mixing of ingredients and provides a mechanical bruising of the straw which accelerates microbial action in decomposition.

The microbiology of mushroom composting attracted relatively little attention since Waksman's early studies, until Hayes (1969) and Imbernon & Leplae (1972) detailed the main groups of organisms that were active. Emphasis was given by Hayes to the dominant groups at different stages of composting. Numbers of aerobic organisms were greatly in excess of the anaerobic organisms reflecting the known requirement for aerobic conditions throughout composting. Also, reflecting the high temperatures which build up in composting heaps, thermophilic and thermotolerant organisms quickly dominate over the organisms which do not flourish above 30°C - the mesophiles (Fig. 1).

In the early stages, the characteristic species are established. The natural mesophile flora subside while populations of the thermophiles and thermotolerants increase. Bacterial populations dominate and their rapid increase in numbers coincides with maximum heat generation and consequently temperature build up. This however only takes place in the central aerobic zones, the outer layers and anaerobic regions at the base of the stack undergo this change following mixing.

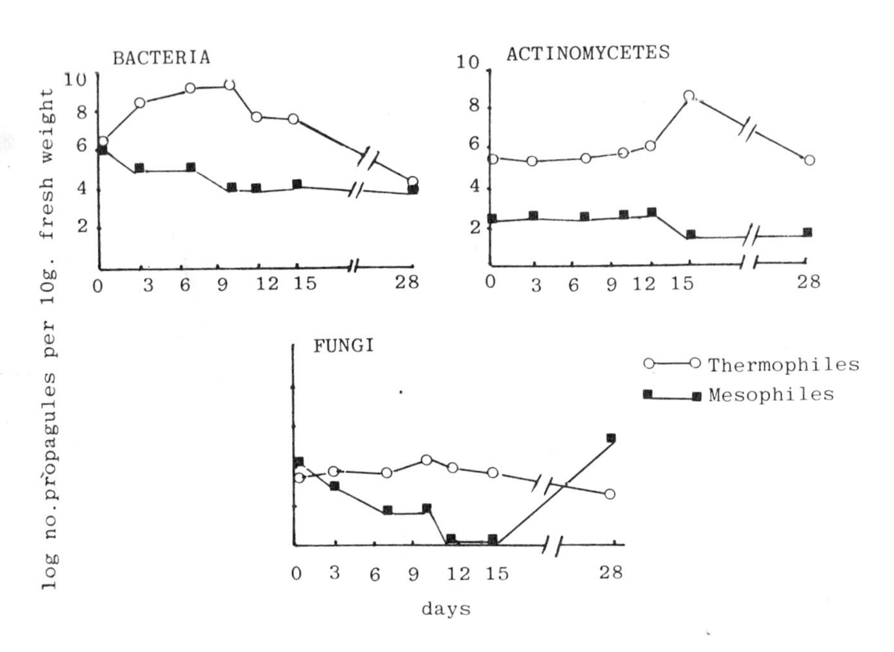

Fig. 1. Changes in the populations of aerobic bacteria, actinomycetes and fungi during composting (day 0-15) and the establishment of *A. bisporus* in compost during incubation (day 16-28) (after Hayes 1969, 1977).

This is followed by a relatively prolonged stage dominated by thermophiles and is characterised by a decline in the numbers of viable bacteria and an increase in the numbers of thermophilic actinomycetes. This is particularly evident during the pasteurisation stage at the end of composting. Fungi, although relatively low in numbers, increase during this stage and are active in the overall decomposition process. In this stage temperatures are, for the most part, within the range 50-60^{0}C but at the end of composting, temperatures decline and the stage is completed when the compost temperature reaches 30-25°C and is inoculated with *A. bisporus*.

Inoculation and Incubation Following the inoculation of the compost substrate with mushroom spawn, the substrate is colonised by the actively growing mycelium of the inoculated species - a mesophile (Fig. 1). Its colonisation is favoured by the natural selectivity of a well decomposed substrate and an incubation temperature of 25°C. The thermophilic and thermotolerant flora decline rapidly during this stage, but the numbers of mesophile bacteria increase. The increase in bacteria is relatively small but is nevertheless significant, while the actinomycete populations are inactive and numbers remain static. Normally, the compost substrate is fully colonised fourteen days following incubation and represents the vegetative or mycelial stage in the life cycle of the fungus.

Induction of fruitbody formation To induce the formation of fruitbodies a second substrate is applied to overlay the colonised compost, a procedure which is commonly referred to as 'casing' mushroom beds. These are cased to a depth of 4 cm, and without this second substrate fruitbody formation will not occur.

The substrate used to 'case' mushroom beds is, relative to the primary compost substrate, a nutritionally dilute medium and normally consists of a moist soil or a wetted peat neutralised with chalk or limestone. Following its application, incubation temperatures are maintained at 25°C in a non-ventilated atmosphere for 7-10 days to encourage the vertical growth of mycelium from the compost to the surface. When the mycelium reaches the surface, the growth atmosphere is ventilated with fresh air and the temperature reduced to 10°C. After a further 3-7 days incubation, the mycelium forms aggregates or initials near to, or on, the surface. A proportion of these develop further to form primordia or what are termed 'pinheads' with a recognisable stalk and cap. These then enlarge and mature to form the characteristic mushroom fruitbody, a reproductive structure consisting of a stipe (stalk), a pileus (cap) which bears the spore bearing tissue the lamellae (gills) on its underside. Mature mushrooms are harvested approximately 21 days after casing the beds, and continue to form in characteristic flushes or breaks at approximately weekly intervals, until such time as the productivity declines to an uneconomic level.

The initiation of fruitbodies in the casing is known to be associated with bacterial activity in this soil layer (Eger, 1961; Urayama, 1969; Hayes *et al.*, 1969; Hayes, 1974) (Fig. 2 and 3). Although the part they play in this

Fig. 2. The action of single isolates of bacteria in the casing soil. Flask A. Inoculated with *A. bisporus* but compost and casing sterilised and maintained sterile. B. as in A, but pure culture of *Pseudomonas putida*, added to the casing soil (Hayes *et al.*, 1969).

important transition from vegetative to reproductive growth is not fully understood, their numbers increase following the application of the casing soil, rising to a maximum level at about 10 days (Hayes, 1974; Hayes & Nair, 1977), a time which coincides with the first stages of initiation, the formation of mycelial aggregates or initials. Their numbers then decline slightly. Hayes *et al.* (1969) and Hayes & Nair (1977) showed that bacterial populations were dominated by the Pseudomonads. Especially relevant to the initiation process were *Pseudomonas putida* and Group IV Pseudomonads. The extent of bacterial activity as shown by their numbers in a peat-chalk casing, was also related to the number of primordia and of mature mushrooms which did form at first harvest. In more recent studies by Cresswell & Hayes (1979) a decline in the numbers of the uppermost surface layers of the compost was observed after casing and this was related to the increase in numbers in the casing soil, suggesting a possible migration of bacteria from the uppermost layer of compost into the casing soil.

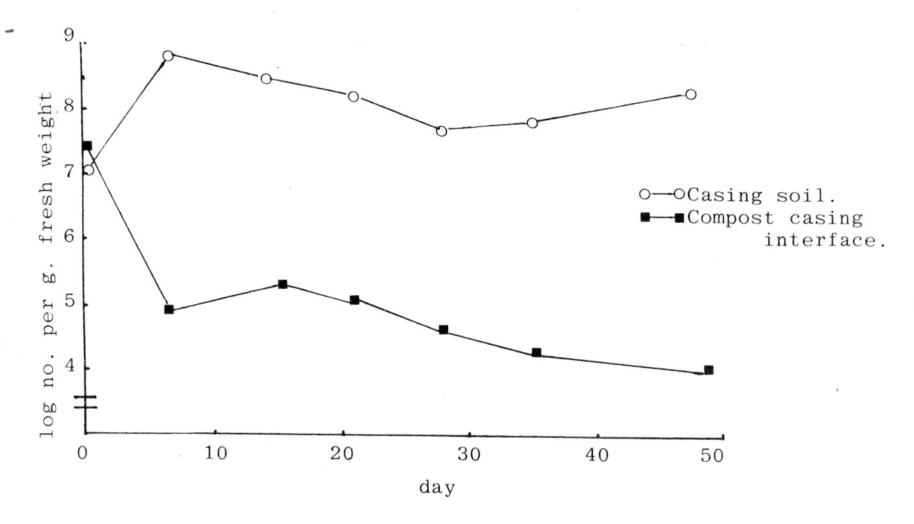

Fig. 3. Changes in the populations of the aerobic bacterial flora in the casing layer and the compost casing interface (after Cresswell & Hayes, 1979). Day 0 = casing applied, day 21 = first harvest, day 48 = third harvest.

Cropping and its termination The two substrates, compost and casing, are of course ideally, fully colonised by the mycelium of the inoculated mushroom from the time of fruit-body induction onwards. However, bacterial activity in the casing layer continues throughout the sequence of flushing of fruitbodies which constitutes the cropping time. This was shown by Cresswell & Hayes (1979) who found a change in the bacterial composition and also a slight increase in numbers, which coincided with the third harvest of fruits (Fig. 3). Although these changes could not be related to any growth feature, they may indirectly be related to the occurrence of organisms pathogenic to the cultivated mushrooms. These include fungus pathogens *Verticillium fungicola* and *Mycogone pernicosa*, and the bacterial pathogens *Pseudomonas tolaasii* and *Ps. agarici*. Their occurrence is common and universal in mushroom cultivation and their frequency increases progressively during the cropping time.

The increase in levels of these organisms must therefore be considered as part of the total involvement of microorganisms in the solid state fermentation and in order to reduce levels of these potentially damaging organisms, beds and all ancillary items, e.g. trays and shelves, the culture house and equipment are disinfected at the termination of the crop.

This is done by the application of heat or chemically by fumigating with methyl bromide, normally *in situ* before disposal of the substrates and the re-use of the trays, shelves and growing house for subsequent crops.

Nutritional and chemical aspects

The minimal nutritional requirements for *A. bisporus* to complete its life cycle have only been defined to the primordian formation stage (Hayes, 1972, 1978), but it is also clear from the studies of Smith & Hayes (1972) who defined a hydroponic experimental system using relatively inert carrier materials, that fruitbody yield is maximal on compost and casing substrates. These substrates are chemically complex and pose particularly difficult problems in technique for chemical analysis. Present knowledge is therefore restricted to carbon and nitrogen nutrition and the role of minerals, trace elements and vitamins and other growth factors are currently under investigation and should provide a basis for a more complete understanding.

The studies of Waksman & Nissen (1932) and Gerrits *et al.* (1965) have shown the primary components of straw, lignin, cellulose and hemicellulose are lost during the composting process and are also utilised by the mushroom *A. bisporus*. During the vegetative stage of growth lignin is utilised, but subsequently to the first harvest cellulose levels decline (Figs. 4 and 5).

Enzymatic studies by Turner *et al.* (1975) have shown that the enzyme laccase, which is responsible for lignin degradation is at a high level during the stage of maximum lignin decline in the compost substrate and cellulose levels increase dramatically at the maturation of the first harvest fruits. Wood & Goodenough (1977) also found a similar pattern.

The enrichment of horse manure and straw mixtures with materials rich in nitrogen and available forms of carbohydrate is universally practised and is important not only to maximise yield, but also to consistently generate an active fermentation (for classification and function of composting supplements see Hayes, 1978). In the absence of any conclusive data relating to yield and the amount of nitrogen in compost, Hayes (1968) stressed the importance of nutrient balance in the computation of mixtures for composting and also the observed relationship between high thermophilic bacterial activity during composting and greater productivity. It was suggested that the build up of bacterial 'biomass' during composting was the key link between growth of *A. bisporus* and a heat producing fermentation. Later work by Stanek (1972) confirmed the importance of bacteria in the nutrition

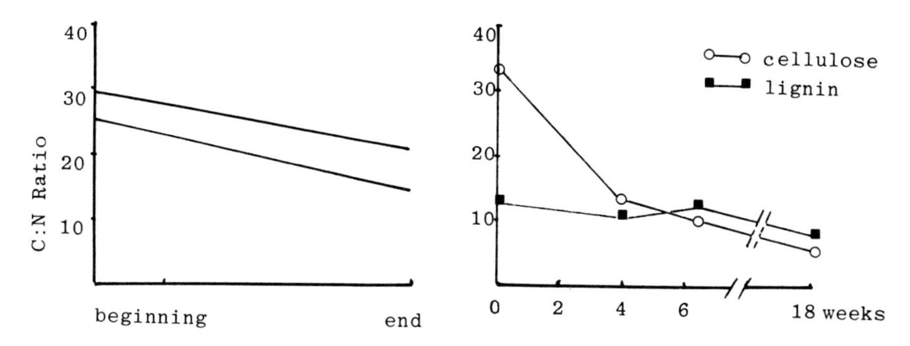

Fig. 4. Changes in C:N ratio during composting (left) and lignin and cellulose levels during composting and growth (right). Composting 0-6 weeks, harvesting 9-18 weeks (after Gerrits *et al.*, 1965).

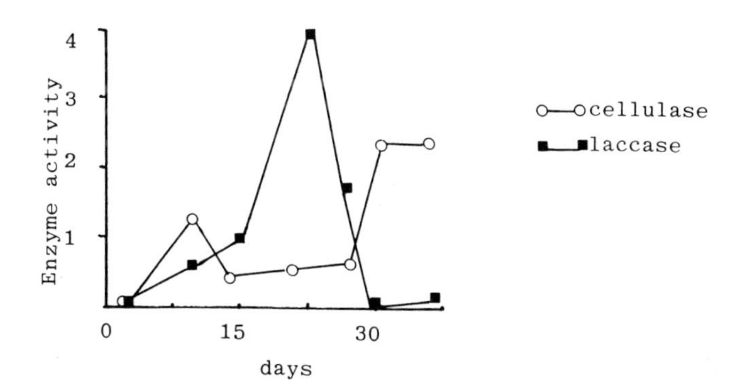

Fig. 5. Changes in the levels of laccase and cellulase in composts during fruitbody initiation and formation (Turner *et al.*, 1975). First harvest = 30 days.

of mushrooms and described a polysaccharide containing nitrogen, separated from a bacterium isolated from compost which stimulated mycelial growth. Eddy & Jacobs (1976) also related the black deposits left on straw after composting, to bacteria and the nutrition of *A. bisporus*.

This link between the chemical components of composts and the ecology of composting has led not only to the adoption of more precise methods (Hayes & Randle 1969, 1970) but also has extended the range of materials that may be successfully exploited for the purposes of mushroom cultivation, hitherto not considered to be suitable substrate ingredients.

Although the casing layer is a nutritionally poor medium, it is not nutritionally inert and the studies of Hayes (1972) have shown that nutritional factors related to the availability and the inhibitory action of certain ions may be relevant to the transition from vegetative to reproductive growth, formation of primordia and fruitbodies. For example, the application of ferrous sulphate to the casing layer in some instances induces a greater number of primordia and maturing fruitbodies (Hayes, 1972; Kaul & Kachroo, 1976; Nair, 1979).

Following the work of Schisler (1967), Schisler & Sinden (1966), and Wardle & Schisler (1969), some interest has been shown in the supplementation of composts subsequent to the establishment of *A. bisporus*, at the time of casing the beds. The addition of meals of cotton-seed and soy bean are the most effective additions at this stage, but more recently a similar effect has been given by encapsulating micro droplets of vegetable oil with a protein coat, following denaturing with formaldehyde (Carrol & Schisler, 1976). This form of nutrient enrichment of substrates already colonised is primarily applicable to some production units in the U.S.A.

Of relevance also to the total fermentation is the application of chemical agents to control the inevitable accumulation of harmful pests and pathogens (see Hussey *et al.*, 1969; Hayes & Nair, 1974). Incorporation of pesticides, such as diazinon, thionazin and malathion in the compost and casing substrates is the most common method of controlling phorids, sciarids and cecids. Fungicides include cufraneb, mancozeb, daconil and benomyl. Chlorine is often applied in water to control bacterial diseases. While the incorporation of such agents may be economically justified, their use in mushroom culture undoubtedly influences the nature of the fermentation - some agents stimulate yield, while others reduce productivity.

Systems and methods

Since mushroom substrates are bulky and very often large quantities are processed at one time, a variety of culture systems are adopted in order, firstly, to optimise the total environment for the various stages and secondly, to permit the exploitation of local resources.

The main characteristics of culture systems are outlined in Fig. 6, but within each system a large number of variations can be found. In composting, for example, a number of innovations have been perfected to give a greater degree of control over the process. The controlled aeration of the traditional composting piles is being practised successfully by the insertion of ducted channels and the provision of fans. Similarly, in what is known as bulk pasteurisation, an entire batch of compost is treated in an insulated tunnel before filling into the growing containers (Italian Patent No. 53693A

Composting

Straw based mixtures arranged in stacks and mechanically mixed and aerated. A mixed fermentation dominated by thermophiles (10-30 days).

Pasteurised in trays or on shelves in a heat room by the injection of steam or methyl bromide (3-10 days).

↓

Inoculation

Incubated on shelves at 25°C (14 days)

↓

Cased and incubated at 16°C for fruitbody formation and harvesting (9-12 weeks)

Incubated in trays at 25°C (14 days)

↓

Cased and transferred to a growing room at 16°C (9-12 weeks)

Pasteurised in bulk or in insulated tunnels. Steam and air forced through the mass (5-7 days).

Inoculation and *incubation* stages can also be included before removal to appropriate container for growing (14-20 days).

↓

Filled and compressed in disposable containers and transferred to a growing room at 16°C (9-12 weeks).

Fig. 6. Outline of procedures in the cultivation of *Agaricus bisporus*.

1971. E. Giodani and B. Francescutti, and Denham, 1977). Air and steam is forced through the entire mass.

A third system, which perhaps elevates mushroom composting into an ideal fermentation system, is one based on a rotating drum principle. This concept was first described by Stoller *et al.* (1937), and in an experimental system was termed Controlled Environment Composting by Randle & Hayes (1972). In this latter system the feasibility of composting substrates and spawning *in situ* before placing into the growing containers were clearly demonstrated. Under commercial conditions, drum composting is only practical for small scale production and a commercial drum composter is illustrated in Fig. 7.

Fig. 7. A revolving drum composter used for the commercial production of composts for *A. bisporus* culture. (Courtesy of J. Baker, Glenian Enterprises, N.S.W., Australia).

Following the inoculation of substrates with spawn, the environmental conditions required for the different stages are critical to growth and the productivity of cultures and since fruitbodies are formed on the surface, substrates are arranged in layers in the growing houses. The two main culture systems which have evolved from the original cave system, where substrates are arranged in ridge beds on the cave floor, are known as the shelf and tray systems. The latter is particularly amenable to mechanical handling and substrates can be moved from specialised rooms for incubation before the application of the casing substrate and then to the growing rooms. Recent improvements to the shelf system have greatly increased the potential of this system. The use of movable nylon nets at the base of the fixed shelves greatly reduces the cost of filling and emptying.

Also, a recent development is the use of disposable containers - the simplest of which is a sack made from polythene. Similarly, spawned substrates can be made into blocks or bales and overwrapped. Growing modules of this kind can be used as a low cost alternative to the shelf or tray system or are sometimes marketed for amateur cultivators. The major advantage of disposable container growing is in disease control. Less rigorous terminal disinfection procedures are required, since they are not recycled and when localised disease outbreaks occur they can be readily isolated from other modules.

Mushrooms are grown in a wide range of buildings, but intensive cultivation, however, requires specialised growing structures commonly referred to simply as growing houses. In recent years the costs of construction have been greatly reduced by structures consisting of a layer of insulation between an inner and outer film of heavy gauge polythene. Such buildings are often more easily disinfected than the traditionally used concrete buildings.

While a range of culture systems exist, ranging from crude to highly sophisticated, a major constraint to the evolution of systems into a continuous process is the relatively slow growth rate, and the characteristic growth pattern and flushing of fruitbodies which protracts harvesting time. Hand harvesting is therefore practised. This involves the selection and systematic cutting of mature fruits by harvesting personnel in the growing houses, but a reverse system can be applied to tray systems of culture, where trays are transferred in stacks to a centralised harvesting area. They are then transferred to moving conveyors with the harvesting personnel cutting on either side of the conveyor as the trays pass along. At the end of the conveyor line, the trays are watered automatically and stacked again for transfer back to the growing house. This system requires a high degree of sophistication in design of growing units and is only rarely applied.

Technically, the mechanical cutting of mushroom fruits is possible. Persson (1972) for example, describes a machine for harvesting and in Holland a movable cutter over the shelf beds has been devised. The harvested product is however only suitable for processing purposes. It appears as if a greater degree of control over the growth of fruitbodies and the elimination of disease organisms which readily spread when fruits are mechanically harvested, is a primary requirement for commercial success.

THE ARTIFICIAL CULTURE OF *VOLVARIELLA VOLVACEA*

Unlike *A. bisporus*, a temperate species, *V. volvacea* is a tropical and sub-tropical species and all cultivated strains have an optimum temperature for growth which varies between 30-35°C. Physiologic differences are also apparent and appear not to be adapted to utilize lignin as a nutrient (Hayes & Lim, 1979). However, straw is used as the primary substrate and its cultivation is confined to tropical and sub-tropical Asian countries where rice straw is used. Also, unlike *A. bisporus*, a casing layer is not required for the induction of fruitbodies, but a light requirement is essential.

The artificial culture of *V. volvacea* has been traditional in the Chinese rural communities for about two centuries and up to comparatively recent times little or no improvements to the traditional methods were attempted. In comparison with *A. bisporus*, the biological nature particularly in regard to its nutrition, has not been extensively studied, but more recent interest in its cultivation suggests that some of the underlying principles of *A. bisporus* culture might be applied to *V. volvacea*, thus improving the prospect of improvements to the traditional method.

Traditional methods

There are a wide range of methods which vary from country to country and from locality to locality. In Hong Kong (Chang, 1972) bundles of rice straw are first thoroughly soaked in water before being piled in layers to about 1m high. Rice bran or distillery wastes are added and the straw stack covered with mats which prevent excessive drying out. This encourages a build-up of temperature within the stack which declines after 3 or 4 days. Inoculation with spawn often takes place at the time of making up the stack and thus following the initial increase in temperature, *V. volvacea* rapidly colonises the straw bed and after a further 14 days primordia and fruitbodies form. There is little or no control over environmental conditions as cultivation is done in the open and therefore subject to climatic variables.

In the traditional Hong Kong method it can be assumed that the initial fermentation is comparable to a composting process. At least there is a visual change in the straw - it turns a characteristic brown colour and is softened by the process. However, in Malaysia no composting was done when rice straw was used for mushroom cultivation (Anon., 1972) and yields were very poor. However, Lim (1976) defined a procedure which included a composting process. Straw is first soaked and then mixed with cattle manure before making into a stack. Composting is allowed to take place for 10 days and turned every 2 or 3 days to maintain aerobic conditions.

The beds are constructed from this composted straw and inoculated following the completion of composting. They are usually arranged in the open but are protected by a mat or plastic covering. About 14 days after spawning unopened mushrooms are harvested and like *A. bisporus*, fruitbodies are formed in sequential breaks. The average yield is about 4.5 kg fresh mushrooms per 100 kg of dry straw, which is double that of the best yields from the traditional methods.

In recent experiments Hayes & Lim (1979) have compared the microbial ecology of composting rice and wheat straw for mushroom cultivation. The species composition is similar and the changes in number of thermophilic and mesophilic bacteria, actinomycetes and fungi follow a comparable pattern. However, rice straw only requires a short time of composting and therefore the thermophilic actinomycetes are less prominent. Thermophilic and mesophilic fungi are, however, in greater number than in composting wheat straw and horse manure mixtures. It is clear that the definition of the microbiological and chemical changes which occur in composting and in the growth of *Volvariella* mushrooms could lead to improved techniques applicable to the conditions in rural communities in Asia. However, outdoor methods are subject to major climatic variables and in most situations can only be grown in appropriate seasons of the year (Fig. 8).

Indoor cultivation

The possibility of more sophisticated techniques of culture have recently been demonstrated by Ho (1972) and Chang (1974). Ho describes a method of cultivation in plastic houses which is applicable to Taiwan while Chang (1974) describes a process which includes the use of cotton waste, a product of the local textile industry in Hong Kong.

In this method, cultivation is done in iron-framed plastic houses in which temperature and aeration can be controlled by the introduction of steam and air by forced ventilation systems.

Composting is done on concrete floors and ingredients include cotton-waste (first grade dropping waste), rice or wheat bran and limestone. This mixture is soaked thoroughly in water and heaped into piles of $1.5m^3$. These are fermented for 4 days and turned once during this time. The compost is filled onto beds arranged in a plastic house at about 10-15 cm deep and steam introduced to raise temperatures to 60^oC and maintained for two hours, after which the room is ventilated and temperature maintained at 53^oC. Within twelve

TRADITIONAL (Hong Kong)	*MODERN*
Rice straw and bran soaked in water and arranged in layers to form a stack which is covered by mats. Temperature increase indicative of thermophilic fermentation (3-4 days).	*Malaysia* Rice straw mixed with cattle manure and arranged in stack and turned to encourage thermophilic fermentation (10 days) *Hong Kong* Cotton waste and bran composted for 4 days and pasteurised on shelf beds (12 hours)
↓	↓
Inoculated with pure culture of *V. volvacea*.	Inoculated
↓	↓
Incubated in the open (temperature 30-40°C; 20-40 days).	Incubated in room with controlled aeration (20-40 days)

Fig. 8. Outline of procedures in the culture of *Volvariella volvacea*.

hours the temperature falls to about 35°C, when the compost is inoculated with spawn and growth and harvesting progresses according to the normal pattern. A four times improvement of yield is claimed for this method and Ho (1972) claims a similar improvement to productivity by cultivating indoors, but also by applying a casing layer, a procedure adopted in *A. bisporus* culture. More recently Hayes & Lim (1978) demonstrated a substantial improvement in primordia formation when cased with peat soil. The reasons for this have yet to be determined.

The success of the modern methods of cultivation in plastic houses and introducing techniques comparable to those employed in *A. bisporus* culture demonstrates a greater potential for this Chinese delicacy. Initial observations indicate that microorganisms are similarly associated with growth. However, because of its requirement for a high temperature, its cultivation is likely to be restricted to tropical areas.

THE CULTURE OF *LENTINUS EDODES*

Lentinus edodes grows naturally on trunk and branches of broad-leaved trees in Japan, China, Indonesia and New Guinea, but is cultivated artificially on the wood of *Quercus* species in Japan where it is the most important cultivated mushroom. Its cultivation was probably introduced to Japan by the Chinese over 300 years ago. It is consumed not only for its unique flavour but also it is believed to be an "elixir of life", a belief which probably stems from claims made during the Chinese Ming Dynasty (1368-1644). Many different substances have been detected in *Lentinus* mushrooms, which may be regarded as medically interesting, e.g. Vitamin D_2 which is induced by sun drying of the fruitbodies, anti-tumour activities of polysaccharides and hypocholesterolemic activity of an adenine derivative. Clinical evidence of any medical benefits are not available (for review see Cochran, 1978).

In Japan, *L. edodes* is known as Shiitake, named after one of its natural host trees, Shii (*Castanopsis cuspidata*), but it is also referred to as the forest mushroom.

The primitive methods of cultivating this species relied entirely on the chance inoculation of dead wood exposed to the natural conditions of the forest where Shiitake was known to grow, but the development of techniques of pure culture inoculum (spawn) has greatly improved the predictability of culture.

Cultivation in forests

The method widely adopted for artificial culture involves

a distinct sequence of procedures, most of which have been perfected over many decades, by "trial and error". The wood of *Quercus serrata* and *Q. acutissima* is favoured for Shiitake cultivation and following the felling of trees in the autumn they remain in the forest for up to 60 days and are then cut into logs about 1 m and 5-15 cm in diameter (Fig. 9).

Dead logs of *Quercus* species

↓

Inoculated with pure culture of *Lentinus edodes*

↓

Logs incubated in forests where temperatures of 24-28°C favour colonisation of wood. Often covered with mats or twigs (1½ years)

↓

Logs transferred to cooler shaded zones of the forest. Temperature range 12-20°C. Fruitbodies formed for 3-7 years in Spring and Autumn

Fig. 9. Outline of the procedures in the culture of Shiitake (*Lentinus edodes*).

Sawdust and rice bran are components of a medium used for making spawn or alternatively wedge-shaped pieces of oak wood are used as the spawn substrates. For inoculation incisions are made into the wood and the inoculum filled into the holes. For one log 20 points of inoculation are normally considered to be sufficient and the inoculation points are waxed in order to prevent evaporation.

The favoured temperature for incubation is between 24°C and 28°C and the conditions during this protracted operation (up to 1½ years) are critical to the success of the culture. The logs are laid in a deliberate manner at an angle to the ground, with cross logs arranged so as to provide ventilation and during this incubation time the logs are usually protected with matting, leaves, tree branches etc. Frequently this procedure is done in selected areas of an evergreen grove. Humidity is also important; excessive moisture encourages the growth of competitor fungi.

After colonisation the logs are transferred to areas of the forest which provide the necessary conditions for the induction of fruitbodies. Shade is necessary and humidity must remain high. Temperature should range between 12° and 20°C. Logs are transferred normally in the winter months and two harvests per annum can be expected, at Spring and Autumn time. Logs will continue to produce Shiitake for 3-7 years and fruitbodies are normally dried artificially before marketing.

Progress towards industrialisation of Shiitake cultivation

The cultivation of Shiitake is an extensive system requiring large areas of land and labour. In recent years mechanical aids have been introduced to allow the expansion of the industry in Japan and a considerable amount of research into the biological nature of Shiitake has led to a better understanding of the factors which govern growth. *Lentinus edodes* grows optimally at pH 3.5 - 4.5 and for the formation of fruits there is a requirement for light, the minimum light intensity was estimated by Ishikawa (1967) to be 10^{-2} to 10^{-4} lux.

The realisation that moisture levels in the wood during incubation are critical to the successful establishment of cultures has greatly increased the efficiency of culture. Living wood after felling seems to resist invasion of Shiitake mycelium and a faster growth of mycelium is obtained when wood is killed by heat or gas treatment, which may be indicative of the action of competitive organisms in natural wood.

During the course of growth, lignin, cellulose and hemicellulose are the primary nutrients utilised, but the mechanisms underlying the switch from vegetative to reproductive growth have not been studied. It is possible that the bark layer may function in a similar manner to the casing layer in *A. bisporus* culture.

Fruitbodies can be readily cultivated on sawdust based media, but industrialisation on a large scale is difficult due to the relatively slow growth rate of this wood saprophyte. However, in recent times greenhouse production has been practised especially in the city suburbs. The logs for greenhouse production are kept in a relatively dry condition after colonisation with *L. edodes* and are soaked for 1 or 2 days in water which stimulates the formation of fruits.

The logs are transferred to a greenhouse which maintains a temperature above 15°C and humidity of the air is maintained at a high level. Shiitake mushrooms grow under these favourable conditions within two weeks.

A PERSPECTIVE FOR THE FUTURE

The production of mushrooms world-wide is increasing at an annual rate of about 10% per annum. It was estimated by Delcaire (1978) that for 1975, 670,000 tons of *A. bisporus*, 130,000 tons of *L. edodes* and 42,000 tons of *V. volvacea* were produced. Although in the past edible mushrooms have been regarded as a luxury food, with the extension of mushroom cultivation to almost all parts of the world and the improvements to artificial techniques of culture, they are now generally available. Unlike those mushrooms gathered from nature, the technique of artificial cultures ensures that cultivated mushrooms when offered for sale are truly edible. Dispelling prejudice, which for centuries has been associated with mushrooms, has helped to increase consumption; they are now regarded as useful foods in modern diets, complementing staple foods. They are often prescribed by dieticians to counter obesity and other syndromes associated with present-day eating habits, and also the health-giving properties of the Shiitake mushroom is an important factor in its consumption in Japan.

If considered more broadly, mushroom cultivation may also be justified on the basis that the saprophytic habit is exploited to convert renewable resources into a quality foodstuff, which may otherwise be regarded as agricultural and industrial wastes.

In this contribution, the importance of other microorganisms in the *A. bisporus* fermentation has been stressed. In common with *V. volvacea* and *L. edodes*, distinct stages in the cultivation process can be identified, viz. substrate preparation, inoculation, incubation, crop harvest and disposal of exhausted substrates. The ecological succession of different groups is crucial to the process of *A. bisporus* culture and it is primarily the understanding of this association with microorganisms together with their associated chemical activities which has led to the industrialisation of the *A. bisporus* culture process. But, even by the adoption of the more modern sophisticated techniques, while economically sound, the efficiency of the process of converting a waste material, such as straw, is low. For example, in *A. bisporus* culture at best 50 kg of fresh mushroom fruit-bodies may be harvested per 100 kg of dry straw, in *Volvariella* culture 25 kg fresh weight can be expected from the most modern techniques but this may be as low as 2 kg in the traditional method. Nevertheless, these conversions are superior to those applicable to some processes currently being advocated to increase the productivity and nutritive value of straw destined for animal feedstuff.

Improvements would be achieved if substrates were exploited to a greater degree. Two possibilities exist. Firstly, other edible mushrooms may be established at stages within the ecological sequence, i.e. before colonisation with *A. bisporus* or after the completion of the colonisation and harvest of *A. bisporus*. Suitable species for this possibility include *Pleurotus oestreatus* and related species, known as Oyster mushrooms, and *Coprinus commatus*, known as the Shaggy Cap. A visualised ecological sequence of dominant ecological types in the total fermentation would be:- Thermophilic bacteria ⟶ Thermophilic actinomycetes ⟶ Mesophilic *Pleurotus* ⟶ Mesophilic *Agaricus* ⟶ Mesophilic *Coprinus* ⟶ disposal.

The second possibility relates to the recycling of substrates. This involves the addition of substances and materials designed to replenish the nutrients utilised by the mushroom in its growth. This will inevitably involve a fermentation comparable to the traditional method of composting before the inoculation of substrates with the cultivated mushroom species. Although the number of cycles per given substrate may be limited, the successful development of recycling could have a profound impact on efficiency both biological and economic.

Of relevance not only to biological efficiency, but also to the agronomic practices of many areas of the world, is the fact that the fermentation process of mushroom culture may be regarded as an enrichment of a renewable plant substrate, which is also partially degraded, during the process of culture. Substrates at the end of their productive cycle are useful organic soil conditioners with a fertilizer value in excess of the traditional organic fertilizer, farmyard manure. This is being exploited to a limited extent in Great Britain to cater for the demands of gardens, allotment societies and sports grounds, but in countries where soil conditions limit the range of agricultural crops that may be grown, the provision of "used" substrates allows for the integration of mushroom culture with other agri- and horticultural practices, to the benefit of the total reproductive cycle from agricultural land.

To what extent can our knowledge of the solid-state fermentation in mushroom cultivation be applied more generally? There is little doubt that within the area of mushroom cultivation, many of the principles and techniques are applicable to other edible mushroom species, hitherto not exploited to any extent. Technically, the least difficult of the edible species to cultivate artificially are the species of *Pleurotus*, popularly known as Oyster mushrooms, *P. oestratus*, *P. florida*, *P. eryngii*, and *P. sajoncaju* (Zadrazil, 1978). They may be grown on wood or straw substrates. These, however, are only favoured in Italy and parts of India. A further disadvantage is that under intensive systems of culture, the mature fruits release their spores which can cause allergic reactions to some persons if inhaled in large numbers (Schulz *et al.*, 1974).

Stropharia raguso-annulata is cultivated to some extent in Eastern Europe by relatively simple techniques on unsupplemented straw substrates while *Kuehneromyces mutabilis*, *Auricularia* species and *Tremella fuciformis* are cultivated on wood substrates in Eastern Europe and Taiwan. Their consumption, however, is restricted to enthusiastic gourmets, but their saprophytic habit could lead to a more widespread acceptance in the future. Prospects for the mass culture of the exotic species of edible fungi, e.g. Ceps (*Boletus edulis*), Truffles (*Tuber* species) and Morels (*Morchella esculenta* and related species) must be considered to be remote at present. An understanding of the edaphic factors which control their occurrence in nature could lead to techniques by which they can be artificially cultivated. Both Ceps and Truffles are known to form mycorrhizal relationships with trees and it is difficult to visualise artificial methods which bear any resemblance to those applied to the true saprophytes (for reviews on other edible species see Chang & Hayes, 1978).

In contrast to liquid fermentation, the solid state fermentation of mushroom culture may be applied successfully to both high and low levels of technology (Fig. 10). In many parts of the developing world, the cultivation of mushrooms is now contributing to an important part in rural development programmes. This can be illustrated by the recent introduction of mushroom cultivation to parts of India, where the climatic conditions are favourable and where mushrooms are a favoured complement to the largely vegetarian diet. On the Western slopes of the Himalayas, for eight months of

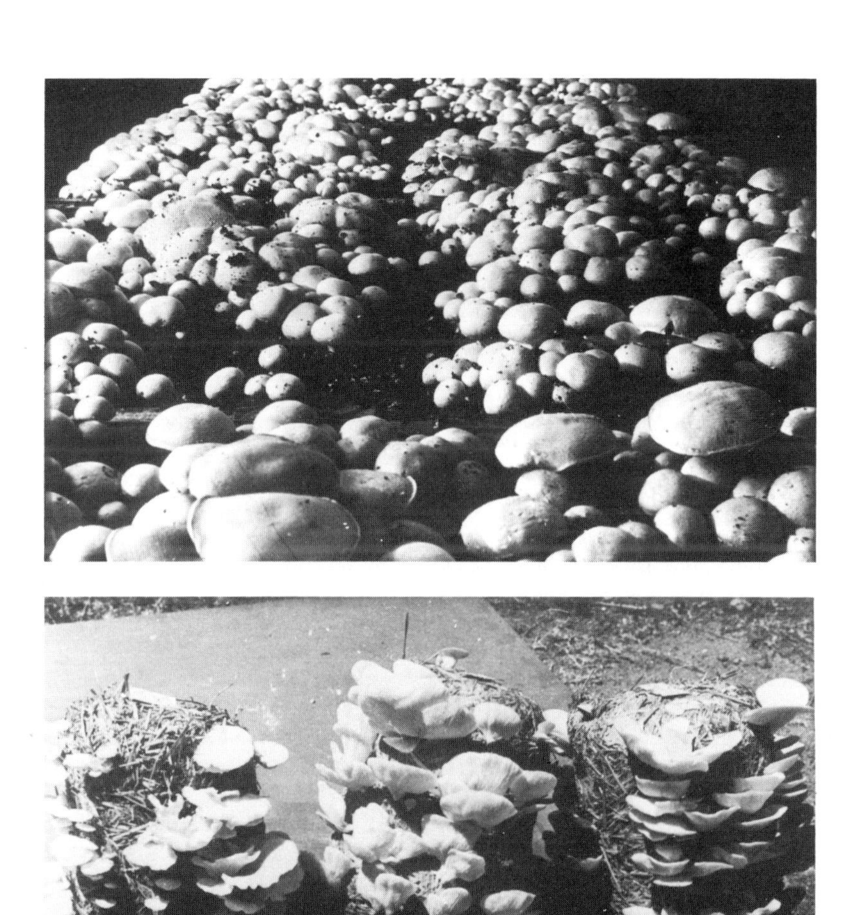

Fig. 10. Contrasting systems of culture. *Agaricus bisporus* (above) is most commonly grown in layers within an environmentally controlled structure. High yields are essential and even mechanical harvesting is feasible. *Pleurotus* species (below) and other species can be cultured satisfactorily on straw bundles in unsophisticated buildings, garden sheds and even dwelling houses. Such simple, low cost systems are particularly applicable to village situations in remote regions of the world.

the year the climate is unsuitable for the cultivation of *A. bisporus* without any air conditioning (heating or cooling). In Himachal Pradesh the production of spawn and compost, relatively sophisticated stages in the culture cycle with high technological inputs, are centred in what are termed "mother" units, which provide a large number of satellite units with the inoculum and substrates for cultivation.

In this way mushrooms are grown in small units such as a mud hut or even in parts of the typical Indian village dwelling. Similarly in Kashmir, many villages are totally involved in mushroom cultivation. Rice straw, which before the introduction of mushroom cultivation was often burnt, is composted by the native villagers, but rely on the supply of spawn from centralised laboratories. Used substrates are an important additive to the agricultural land and thus improve the total productivity of the agricultural system.

Internationally aided development projects provide not only research and advisory inputs but also extensive programmes of education into the basic principles of the mushroom fermentation are undertaken, even at middle school level. In these regions thefore mushroom cultivation assumes an equal status to the cultivation of rice, the by-product of which is vital to mushroom culture and clearly demonstrates how macro- and micro-culture can be integrated to the benefit of rural communities.

The saprophytic habit of edible mushrooms and the unlimited supply of exploitable materials in all parts of the world, provides a theoretical basis for an unlimited potential for the production of this quality food. Its future growth, however, is dependent on a continued growth in consumption. Although a unique food particularly in regard to its texture and flavour, increased consumption will largely depend on costs relative to other foods. Modern techniques which are applied to the industrialised Western Countries elevate mushroom cultivation to an industrial process producing acceptable and edible biomass. The solid-state fermentation of mushroom culture is largely a natural process and can also be applied to cottage scale enterprises, with varying levels of technological and production costs.

REFERENCES

ANON. (1972). Cultivation of padi straw mushroom. *Ministry of Agriculture and Fisheries*, Malaysia Leaflet No. 47.

CARROL, A.D. & SCHISLER, L.C. (1976). A delayed release nutrient supplement for mushroom culture. *Mushroom News* 7,4.

CHANG, S.T. (1972). *The Chinese Mushroom*. The Chinese University of Hong Kong.
CHANG, S.T. & HAYES, W.A. (1978). *Biology and Cultivation of Edible Mushrooms*. Academic Press, New York.
COCHRAN, K.W. (1978). Medical effects. In *Biology and Cultivation of Edible Mushrooms*. Eds. S.L. Chang and W.A. Hayes Academic Press, New York.
CRESWELL, P.M. & HAYES, W.A. (1979). Further investigations on the bacterial ecology of the casing layer. *Mushroom Science* II (In Press).
DELCAIRE, J.R. (1978). Economics of cultivated mushrooms. In *Biology and Cultivation of Edible Mushrooms*. Eds. S.T. Chang and W.A. Hayes. Academic Press, New York.
DENHAM, T.G. (1977). The 3-phase-1 system - past, present and future roles in the new development of mushroom composting. In *Composting, Proceedings of Aston Seminar in Mushroom Science* 2, 67.
EDDY, B.P. & JACOBS, L. (1976). Mushroom compost as a source of food for *Agaricus bisporus*. *The Mushroom Journal* 38, 56.
EGER, C. (1961). Untersuchungen eiber die Funktion der Deckschicht bei der Fruchtkoerperbildung des Kulturchampignons. *Archiv für Mikrobiologie* 39, 313.
GERRITS, J.P.G., BELS-KONING, H.C. & MULLER, F.M. (1965). Changes in compost constituents during composting, pasteurisation and cropping. *Mushroom Science* 6, 225.
HAYES, W.A. (1969). Microbiological changes in composting wheat straw/horse manure mixtures. *Mushroom Science* 7, 173.
HAYES, W.A. (1972). Nutritional factors in relation to mushroom production. *Mushroom Science* 8, 663.
HAYES, W.A. (1974). Microbiological activity in the casing layer and its relation to productivity and disease control. In *The Casing Layer, Proceedings Aston Seminar in Mushroom Science* 1, 37.
HAYES, W.A. (1977). Mushroom nutrition and the role of microorganisms in composting. In *Composting. Proceedings of Aston Seminar in Mushroom Science* 2, 1.
HAYES, W.A. (1978). *Agaricus bisporus* Biological Nature. In *Biology and Cultivation of Edible Fungi*. Eds. S.T. Chang and W.A. Hayes. Academic Press, New York.
HAYES, W.A. & LIM, W.C. (1979). Wheat and rice straw composts for mushroom production. In *Straw Decay*, Ed. E. Grossbard. John Wiley, Chichester.
HAYES, W.A. & NAIR, N.G. (1974). The cultivation of *Agaricus bisporus* and other edible mushrooms. In *The Filamentous Fungi*, Ed. J.E. Smith and D.R. Berry, Volume 1, p. 212. Edward Arnold, London.

HAYES, W.A. & RANDLE, P.E. (1969). Use of molasses as an ingredient of wheat straw mixtures used for the preparation of mushroom composts. *Report, Glasshouse Crops Research Institute,* 1968, 142.

HAYES, W.A. & RANDLE, P.E. (1970). An alternative method of preparing mushroom compost, using methyl bromide as a pasteurising agent. *Annual Report, Glasshouse Crops Research Institute* 1969, 166.

HAYES, W.A., RANDLE, P.E. & LAST, F.T. (1969). The nature of the microbial stimulus affecting sporophore formation in *Agaricus bisporus* (Lange) Sing. *Annals Applied Biology* 64, 177.

HO, MING-SHU (1972). Straw mushroom cultivation in plastic houses. *Mushroom Science* 8, 257.

HUME, D.P. & HAYES, W.A. (1972). The production of fruitbody primordia in *Agaricus bisporus* (Lange) Sing. on agar media. *Mushroom Science* 8, 527.

HUSSEY, N.W., READ, W.H. & HESLING, J.T. (1969). *The Pests of Protected Cultivation.* Edward Arnold, London.

IMBERNON, M. & LEPLAE, M. (1972). Microbiologie des substrates destines a la culture du champignon de couche. *Mushroom Science* 8, 363.

ISHIKAWA, H. (1967). Physiological and ecological studies on *Lentinus edodes* (Berk) Sing. *Journal Agricultural Laboratory* (Gen.) 8, 1.

KAUL, T.N. & KACHROO, J.L. (1976). Use of ferrous sulphate for mushroom farms in Kashmir. *The Mushroom Journal* 43, 214.

LAMBERT, E.B. (1941). Studies on the preparation of mushroom compost. *Journal Agricultural Research* 62, 415.

LAMBERT, E.B. & DAVIS, A.C. (1934). Distribution of oxygen and carbon dioxide in mushroom compost heaps as affecting microbial thermogenesis, acidity and moisture therein. *Journal Agricultural Research* 48, 587.

LIM, W.C. (1976). An improved method of cultivation of *Volvariella volvacea* using padi straw. *Proceedings Symposium on soil microbiology and plant nutrition. University of Malaysia Press* Pre-Print.

NAIR, N.G. (1978). A report on mushroom work. *Australian Mushroom Growers Annual* 1, 34.

PERSSON, S.P.E. (1972). Mechanical harvesting of mushrooms and its implications. *Mushroom Science* 8, 551.

PIZER, N.H. (1937). Investigations into the environment and nutrition of the cultivated mushroom. *Psalliota campestris.* I. Some properties of composts in relation to the growth of mycelium. *Journal Agricultural Science* 27, 349.

PIZER, N.H. & THOMPSON, A.J. (1938). Investigations into the environment and nutrition of the cultivated mushroom *Psalliota campestris*. II. The effect of calcium and phosphate on growth and productivity. *Journal Agricultural Science* 28, 604.

RANDLE, P.E. & HAYES, W.A. (1972). Progress in experimentation on the efficiency of composting and composts. *Mushroom Science* 8, 789.

SCHISLER, L.C. (1967). Stimulation of yield in the cultivated mushroom by vegetable oils. *Applied Microbiology* 15, 844.

SCHISLER, L.C. & SINDEN, J.W. (1966). Nutrient supplementation of mushroom compost at casing. Vegetable Oils. *Canadian Journal Botany* 44, 1063.

SCHULZ, K.H., FELTEN, H. & HAUSEN, B.M. (1974). Allergy to the spores of *Pleurotus florida*. *The Lancet* 5, 1974.

SMITH, J.F. & HAYES, W.A. (1972). Use of autoclaved substrates in nutritional investigations on the cultivated mushroom. *Mushroom Science* 8, 355.

STANEK, M. (1972). Microorganisms inhabiting mushroom compost during fermentation. *Mushroom Science* 8, 797.

STOLLER, B.B., SMITH, F.B. & BROWN, P.E. (1937). A mechanical apparatus for the rapid high temperature microbial decomposition of fibrous cellulosic materials in the preparation of composts for mushroom culture. *Journal American Society Agronomy* 29, 717.

TURNER, E.M. (1974). Phenoloxidase activity in relation to substrate and development stage in the mushroom *Agaricus bisporus*. *Transactions British Mycological Society* 63, 541.

TURNER, E.M., WRIGHT, M., WARD, D., OSBORNE, D.J. & SELF, R. Production of ethylene and other volatiles and changes in cellulase and laccase activities during the life cycle of the cultivated mushroom *Agaricus bisporus*. *Journal General Microbiology* 91, 167.

URAYAMA, T. (1961). Stimulative effect of certain specific bacteria upon mycelial growth and fruitbody formation of *Agaricus bisporus* (Lang) Sing. *Botanical Magazine* (Tokyo) 74, 56.

WAKSMAN, S.A. & NISSON, W. (1932). On the nutrition of the cultivated mushroom *Agaricus campestris* and the chemical changes brought about by this organism in the manure compost. *American Journal Botany* 19, 514.

WARDLE, K.S. & SCHISLER, L.C. (1969). The effects of various lipids on growth of mycelium of *Agaricus bisporus*. *Mycologia* 61, 305.

WOOD, D.A. & GOODENOUGH, P.W. (1977). Fruiting of *Agaricus bisporus*. Changes in extracellular enzyme activities during growth and fruiting. *Archiv für Mikrobiologie* 114, 164.

ZADRAZIL, F. (1978). Cultivation of *Pleurotus*. In *Biology and Cultivation of Edible Mushrooms*. Eds. S.T. Chang and W.A. Hayes. Academic Press, New York.

DEVELOPMENTS IN INDUSTRIAL FUNGAL BIOTECHNOLOGY

B. Kristiansen[1] and J.D. Bu'Lock[2]

[1]*Department of Applied Microbiology, University of Strathclyde, Glasgow G1 1XW*

[2]*Microbial Chemistry Laboratory, Department of Chemistry, University of Manchester, Manchester M13 9PL*

INTRODUCTION

This survey is about fermentation related processes - the traditional aspect of biotechnology. Over the past ten years we have seen considerable growth in biotechnology. In addition to advances made in traditional areas, biotechnology has found application in a range of new fields. The fact that large state grants have been made available in a number of countries illustrates that biotechnology has come of age.

Topics covered in this chapter include production of organic acids, pharmaceuticals and enzymes and waste utilisation. Two sections are devoted to the production of 'new' products which are currently the subject of considerable interest, *viz*. microbial polysaccharides and insecticides. Finally there is a section on trends in fermenter design. The object is to scan the field to give a measure of the strength of the fermentation industry. It may be considered that a survey on the fermentation industry is severely restricted by eliminating bacterial processes. Hopefully it will be demonstrated that fungi are industrially important microorganisms which sometimes are not given the attention they merit.

ORGANIC ACIDS

A number of surveys on organic acids produced by fermentation have been published recently (Miall, 1978; Lockwood, 1974). Being limited to organic acids the reviews are much richer in detail and the reader is advised to consult these if further information is required. The object in this section is to describe the industry, its technology and its future.

Citric Acid

Citric acid is the major organic acid produced by fermentation. Synthetic production is feasible but not competitive. The world production has been estimated at approximately 300,000 tons/year. About 60% of the citric acid finds application in the food industry, other major citric acid users are the pharmaceutical and chemical industries. As any student of microbiology will know, production of citric acid has been researched in great length. In spite of this, however, the large yields of citric acid obtained are based on experience and know-how rather than scientific understanding.

Citric acid is produced in both surface and submerged culture, using strains of *Aspergillus niger*. In the surfacc culture, the citric acid producing mycelium is grown on molasses (or other suitable substrates) in shallow trays. These are kept in stacks in constant temperature rooms. Oxygen is supplied by blowing air over the trays. The air is often used to control the temperature. Control of the fermentation, however, is obviously virtually impossible. The submerged culture process is normally carried out in stirred tank fermenters. However, the oxygen demand is sufficiently low to employ tower fermenters, reducing the installation and operating costs. Recently a comparison based on West German data was made between the surface and submerged process (Schierholt, 1977). The findings are summarised in Table 1.

Table 1. Comparison of surface *vs* submerged fermentation costs. (Assumed output of citric acid 72 tonnes /day (Schierholt, 1977).

	Surface	Submerged
Absolute costs (10^6 DM/Yr)	28	94
Cost breakdown (%)		
Buildings (at 3.3%)	7	2
Plant (at 6.7%)	6	8
Materials	38	37
Energy	3	30
Labour	46	23

Only the fermentation step is covered in the data. Certain aspects of the analysis may be criticised, but it illustrates that the 'ancient' surface culture may still represent economically viable technology. The choice between surface and submerged depends on a number of factors. Low grade molasses tend to give a higher product yield in surface culture and for small scale operation (<5 tonnes/year) it is claimed that this is the more economical method. It must be noted, however, that a much higher product formation rate is obtained in a submerged process.

The mould morphology obtained in the submerged culture has been the subject of considerable discussion. The general consensus seems to be that the formation of small dense pellets, which have been described in detail (Clark, 1962), is a prerequisite for the formation of citric acid. This may not be wholly correct, however, comparing a pellet producing tower fermentation with a stirred tank process in which pellets were not formed, Clark & Lentz (1963) observed no reduction in citric acid yield. It may well be that the formation of pellets is a result of the same environmental conditions which favour formation of citric acid. It is desirable, however, to produce pellets for process recovery consideration. The dense pellets are much more easily separated and washed than filamentous mycelium.

Continuous production has been shown to be feasible (Kristiansen & Sinclair, 1979). However, an industry which does not find it necessary to change from surface to submerged culture is unlikely to consider continuous production in the foreseeable future.

The citric acid industry has recently, however, been subject to an exciting development. With the interest in hydrocarbon as raw materials, fermentation processes were developed for producing citric acid from n-alkanes, using yeasts. For economic/social/political reasons citric acid hydrocarbons were not allowed in food, thus putting an end to these processes. They were subsequently modified to accept sugars as a raw material. The major citric acid procuders are now operating in addition to the traditional production with *A. niger*, molasses based yeasts processes. This opens up the possibility of reducing the fermentation time (up to 50%). In addition, the yeast, normally strains of *Candida*, are more tolerant to trace metals than *A. niger*, rendering the expensive substrate pretreatment steps unnecessary. At the moment, yields of citric acid in the yeast processes are relatively low and it is necessary to operate with lower sugar concentration. This will be overcome, however, and we may well find that a substantial proportion of the classic "citric acid by *A. niger*" will be replaced by processes employing yeasts.

Itaconic acid

Itaconic acids can be produced via three different routes, namely pyrolysis of citric acid, decarboxylation of aconitic acid and fermentation. The first is not competitive, but the second offers an alternative to the fermentation process by which the bulk of the itaconic acid is produced. Itaconic acid is one of the main acids in cane juice. Its removal from molasses is often required for the quality of the final product in sugar processing. The acid is isolated in a relatively simple separation process and represents a cheap raw material for the itaconic acid.

In the fermentation process, using strains of *Aspergillus terreus*, both surface and submerged culture are employed, with the latter being much more efficient. Itaconic acid is used in the plastics industry, but the demand is relatively low as its position is not very competitive. It is, therefore, produced on a small scale and this is the main reason why surface culture is still being used. In general, its situation is very similar to citric acid. Both require relatively expensive substrates and process recovery is complex and costly. A reduction in cost of raw material would make itaconic acid more competitive. It would also benefit from an increase in scale but the present demand does not justify this. A two-stage continuous process has been established (Ryder & Strong, 1967), but is currently not in operation.

Gluconic acid

This acid is produced in submerged culture with strains of *Aspergillus niger*. The raw material is normally glucose, supplemented with a cheap nitrogen source. The fermentation time is approximately 36 h, which is very short for a mould process. The mode of product formation lends itself to continuous culture and a Japanese continuous process has been developed with a productivity of about twice that of the batch process.

Gluconic acid can also be produced enzymatically.

$$\text{glucose} + O_2 \xrightarrow[\text{oxidase}]{\text{glucose}} \text{gluconate} + H_2O_2$$

$$H_2O_2 \xrightarrow{\text{catalase}} H_2O + \tfrac{1}{2}O_2$$

Air can be used to supply the oxygen, and a crude enzyme mixture, already available commercially, is sufficient to carry out the reactions. The system lends itself to immobilization and this may well replace the fermentation process.

However, long term stability of the catalase compound is at present the limiting factor. Presently, synthetic gluconic acid is competing with the fermentation product, making it imperative to have an efficient process.

Malic acid

Malic acid can be produced in submerged culture using strains of *Aspergillus* on glucose. Yields are low, however, and most commercial malic acid is produced synthetically. Recently an alternative method has been introduced in which the acid is produced by fixed enzymes.

Fumaric acid

This acid has been produced commercially in submerged culture using strains of *Rhizopus* with glucose as raw material. The fermentation time is about 5 days, with yields of up to 70%. Fumaric acid is used in the food and pharmaceutical industries. The greatest potential for production of fumaric acid lies in the conversion to malic acid for use in pharmaceuticals. A hydrocarbon based process, using *Candida* has been established, but this is strictly limited to non-food application. Similarly to most other organic acids the demand for fumaric acid will increase if the production cost can be decreased, e.g. finding a cheaper raw material.

Conclusion

In general, the fermentation process is not the main obstacle to an economically viable process. As most of the acids are used in the food industry, substrate costs are necessarily high. Secondly, the product recovery is complex and too costly for a low value product. Improvements in this stage of operation, however, will have a minor impact on the overall process compared to reduction in substrate or product recovery costs.

PHARMACEUTICALS

It was suggested a few years ago that industrial microbiology was limited to production of pharmaceutical and citric acid (with apologies to the brewer). Although this is no longer true, the pharmaceutical industry has offered a considerable contribution to the development of biotechnology. The high growth rate experienced by the industry, with new processes and technological developments flourishing, belongs to the past, however, The risk rate is high, some three quarters of new introductions fail to achieve any significant success. In addition, the cost of developing a new product has increased considerably. The cost of demonstrating the safety and acceptability of products is prohibitive, only the most promising products are carried beyond the laboratory stage. The pharmaceutical industry is, therefore, currently more concerned with new products rather than new processes. The biotechnology is to a large extent dominated by strains of *Penicillium*, *Cephalosporium* and *Streptomyces*, growing in a stirred tank fermenter with advanced process control.

Penicillins

Penicillin is produced in submerged culture, using strains of *Penicillium chrysogenum*. In the early days of the fermentation, surface culture using *P. notatum* were employed. With the introduction of submerged culture, *P. chrysogenum* was found to be a better producer, giving yields of 300 units/ml. Currently, yields up to 25,000 units/ml have been reported and even higher yields can be expected. The increase has been achieved through strain improvement, medium development and proper control of the fermentation process. The addition of a precursor, notably phenylacetic acid, is essential for high yields of penicillin. Controlled continuous addition is necessary to prevent the mould metabolising the precursor. The fermentation is carried out in a batch or fed-batch operation, with the latter taking over from the traditional batch process, increasing the efficiency of the fermentation.

The fermentation product, penicillin G, is rapidly becoming a 'mere' raw material for semisynthetic penicillins. Appropriate side-chains are incorporated into the penicillin molecule either synthetically or enzymatically, producing a more potent product (Moss, 1977). The starting point for most of these 'new' products is 6APA, which is produced by the action of microbial penicillin acylase on penicillin G. The enzyme penicillin acylase is normally obtained from particular strains of *E. coli*, and it is technically feasible to employ it in an immobilized reactor.

The penicillin process serves to illustrate the production of pharmaceuticals in general. New strains and media are continually evaluated to increase the yield in the fermentation. The products may be modified, enzymatically or synthetically to give a more potent antibiotic.

Gibberellins (gibberellic acid)

The gibberellins are produced by *Gibberella fujikoroi*. The most important product is gibberellin A3 (gibberellic acid), but gibberellins A4 and A7 have also found some commercial application. The fermentation is well known, being run as a fed-batch process with an extended stationary stage in which the gibberellic acid is produced. A continuous culture process has been established (Bu'Lock, Detroy, Hostalek & Munion-al-Shakan, 1974) without offering any advantage over the commercial process. The gibberellins are potent plant hormomes, regulating plant growth. Industrial application is in fact limited to horticulture.

ENZYMES

The enzyme industry has experienced considerable growth over the past two decades. The take-off point was represented by the introduction of bacterial amylase and protease in detergents. In terms of tonnage these two enzymes still lead the field (Aunstrup, 1977) but other enzyme preparations may be more valuable. At present the enzyme industry is still relatively small, but active. The main R & D work is to find new industrial applications for existing enzymes. In fact, producing industrial enzymes in the fermenter is relatively simple, the product recovery is the main obstacle to higher yields. Losses of 50% of enzyme activity is often encountered during product recovery. The bulk of commercial enzymes are of bacterial origin. There are, however, some important fungal enzymes, both in terms of tonnage and value. The production of the major fungal enzymes are described below. Further details can be obtained in a number of reviews recently published (Blain, 1975; Barbesgaard, 1977).

Cellulase

Cellulase is produced using strains of *Trichoderma*. The production of this enzyme is a much researched process and details of the process have been published (Sternberg, 1974). Application of cellulase from *A. niger* has also been reported (Bolaski & Gallantin, 1962). Production in submerged culture is now the favoured process. A traditional semi-solid process is used in Japan to a limited extent. The claim to fame of the enzyme lies in the ability to degrade cellulose. Considerable effort is currently being channelled into the production of glucose from cellulose. Two bottlenecks are preventing the process from becoming economically viable. Substantial pretreatment of cellulose prior to saccharification is required and the cost of cellulase production is too high. Work is being carried out to improve the cellulase yield in the fermentation. Both the medium composition and in particular strain improvement is being considered.

Cellulase has also found limited application in the food industry, upgrading dehydrated fruits and vegetables. It is used as an additive to animal feed to increase cellular digestion.

Another potential use of cellulase lies in secondary oil recovery in a process termed micellar flooding. This involves pumping cellulose down existing oil wells to force out remaining oil. Cellulase is added to reduce the viscosity of the cellulose slurry. Presently, the production cost is too high for the oil recovered this way. With the present rate of increase in the oil price, however, secondary recovery of the considerable amount of oil left in existing wells will become economically viable.

It is clear that cellulase has considerable potential, justifying the efforts currently put into its production. This is also illustrated in the chapter "Fuels from Biomass".

α-Amylase

Fungal α-amylase is produced by *A. niger* and *A. flavus oryzae* with the latter being favoured by the industry. As the *A. flavus oryzae* group may on certain occasions be producers of aflatoxin, tests for this compound are normally carried out although the α-amylase producer is not known to produce aflatoxin (Barbesgaard, 1977). Submerged culture is preferred in the West. In the fermentation it is imperative to avoid build-up of glucose which will depress enzyme formation. This required careful balancing of the medium composition. The enzyme is produced on a large scale in Japan using an adaptation of the Koji fermentation. This involves cultivating the microorganisms on the surface of moist bran, or rice, in trays or rotating drums.

α-amylase hydrolyses starch molecules into maltose and maltotriose. It is used extensively in the baking industry where wheat deficient in amylase will be fortified with the fungal enzyme preparation. It is also used in the production of 'higher conversion' syrups. A mixture of α-amylase and amyloglucosidase is added to acidified syrup in a saccharification step to produce a mixture of glucose and maltose, used in confectionary. The enzyme is used to a limited extent by the brewing industry to hydrolyse the starch in malted barley, supplementing the α-amylase which occurs naturally in the barley. Fungal α-amylase is losing the battle with the bacterial enzyme. At present, approximately five kg of bacterial α-amylase are sold per kg of α-amylase of fungal origin. This is a reflection of the increasing sale of the bacterial enzyme, however.

Amyloglucosidase

A number of fungi are known to produce amyloglucosidase, with strains of *Aspergillus* in the forefront. As with other enzymes, submerged culture, using stirred tank fermenters, is the preferred mode of production in the West. Surface culture is still popular in the East. A Japanese surface culture process with *Rhizopus* has been developed for commercial production (Underkofler, 1968). In the deep tank process *A. niger*, *A. oryzae*, *A. foetidus* or *A. amawori* are used with good quality maize as the main raw material. Through strain selection and careful process control the fermentation has become very efficient, with yields above 1 gram active enzyme per litre. The profit margin in amyloglucosidase production is very low and high yields are essential for the process to be viable. To paraphrase a spokesman for the industry "amyloglucosidase is now so cheap you could throw it away". The annual production is about 300 tonnes of pure enzyme (Aunstrup, 1977).

Amyloglucosidase has considerable commercial applications. It is primarily used in hydrolysis of starch. Previously the starch industry employed acid hydrolysis to produce dextrose from starch. Enzyme hydrolysis will increase the yield and produce a higher quality product. The enzyme is also used, together with fungal α-amylase, in the production of 'high conversion' syrups as described above, and in the brewing industry to degrade dextrin. Addition prior to the fermentation will increase the alcohol content and after the fermentation will lead to a sweetened beer.

Pectinase

Pectinase is formed by many microorganisms, but only *A. niger* has been successfully explored on a large scale. Commercial pectinase contains a number of pectic enzymes. For this reason surface culture, growing the fungus on bran, has been adopted by the industry, as this method will produce the required enzyme mixture. Submerged cultivation tends to reduce the number of pectic enzymes produced and a number of fermentation products must be added to produce the mixture (Barbesgaard, 1977). Major applications of pectinase are in the clarification of fruit juices and wines and as gelling agent in jams. The world production of pectinase was estimated to be 20 tonnes pure enzyme in 1977.

Lactase

Lactase is produced commercially on a limited scale. The

The source organism is determined by enzyme application. The dairy industry employs lactase from *Saccharomyces fragilis*, hydrolysing the lactose in condensed milk to avoid lactose crystallisation. The baking industry on the other hand, prefers lactase from *A. niger* or *A. foetidus*. A greater potential application of lactase lies in the breakdown of waste whey, and hydrolysis of lactose in milk. Lactose intolerance is quite common in Asia and parts of Africa and treatment with lactase would overcome this problem. Utilisation of waste whey is described in a later section.

Immobilized enzymes

The bulk of industrially produced enzymes find application in food and related industries. Traditionally most processes involving enzymes are batchwise processes at the end of which the enzyme is discarded. This restricts the enzyme industry to the production of low cost enzyme preparations. The immobilization of enzymes is an attractive concept for a number of reasons. Retaining the enzyme is less wasteful and with a fixed enzyme reactor the enzymatic process can be run continuously. It is also attractive because it may allow more expensive enzymes to be used, opening up new fields of enzyme application. This is illustrated by the development of enzyme electrodes for analysis and process control (Barker & Somers, 1978). It is felt that, paradoxically, the future for the enzyme industry lies in immobilized enzymes. To survey the field of immobilized enzymes is beyond the scope of this review, the field has been illustrated in detail elsewhere, however (Messing, 1977; Olson & Cooney, 1974).

BIOLOGICAL INSECTICIDES

Biological insecticides are designed to behave as a natural control system, being relatively selective in their action. Synthetic pest control agents may carry undesirable side-effects in addition to their designed control action. They may, therefore, be harmful for the general ecology. The most widely produced and marketed biological control agents for insects are based on preparations of bacteria of the *Bacillus thuringiensis* group. However, the use of fungal pathogens and their derived toxins is now technically feasible and several agents, have been tested. Spore production is essential if an infective fungal preparation is to be obtained. For this reason, surface rather than submerged culture is employed, the final product is usually an aqueous spore suspension. Three fungal species,

Beauveria bassiana, *Metarrhizium anisopliae* and *Acrostalagmus aphidum* have been cultivated on a moderate scale. Although many species of fungi are pathogenic for a variety of insect pests, there has been limited success with fungal pesticides in this field.

Biological insecticides is an area on the very fringe of biotechnology, primarily because the areas of application are not directly familiar to biotechnologists. It serves to illustrate, however, how the field of biotechnology is expanding, offering notable contributions to areas traditionally quite separate from the original concept of biotechnology.

POLYSACCHARIDES

Microbial polysaccharides are one of the 'new' products being explored by the fermentation industry. Polysaccharides are normal constituents of the microbial cell. The location of polysaccharides, either within the cell or external to it, provides a basis for their classification. They fall into three categories, intracellular, structural cell components and extracellular polysaccharides. Microorganisms which produce the extracellular type have the greatest potential for industrial development mainly because of the relative ease of product recovery. Polysaccharides are rapidly emerging as important polymers which have novel properties, having already found application in the food, pharmaceutical and oil recovery industries. The two main commercially produced microbial polysaccharides, xanthan and dextran, are bacterial products. Industrially important fungal polysaccharides are at present limited to pullulan and scleroglucan.

Pullulan

Pullulan is produced in submerged culture by *Aureobasidium pullulans*. The fermentation lasts 5-7 days, careful control of pH, particularly during the initial stages, is important for an efficient process. In common with most other polysaccharide fermentations, pullulan is produced during the stationary phase, which is promoted by the depletion of nitrogen in the growth medium. Production in continuous culture under nitrogen limiting conditions has been achieved (Catley & Kelly, 1975). The dilution rate was too low, however, to explore the productivity advantages which may be offered by continuous culture.

It is claimed that pullulan is potentially one of the most important microbial polysaccharides to be produced industrially, with some small scale plants already in operation. Plans to construct a commercial plant with a capacity of 5,000 -

10,000 tonnes has been described (Yuen, 1974). The polysaccharide is produced in a batch process with a glucose syrup as a raw material. Even larger plants are envisaged. The properties and use of pullulan have been reviewed (Zajic & Leduy, 1976).

Scleroglucan

A glucan produced by *Sclerotium glucanium* was developed for potential commercial use in the early 1960's. The fermentation lasts 4-6 days, depending to some extent on the temperature and the carbon source. An important property of scleroglucan is its pseudoplasticity being to a large extent unaffected by pH, temperature and the presence of various salts; the latter in particular affecting many other polysaccharides. Commercial production is well established, although the operation is still on a relatively small scale. Little information is available on the production, however. This is normally an indication that the companies concerned are making a healthy profit.

Commercial production

There is considerable activity in the field of microbial polysaccharides. Many companies in Europe and USA have entered the field, either as an expansion of existing technology or for diversification reasons. Production of microbial polysaccharides represents a considerable challenge to 'traditional' biotechnologists. During the latter half of the batch fermentation the broth becomes highly viscous, exhibiting non-Newtonian pseudoplastic behaviour. The conventional flat-bladed turbine employed in stirred tank fermenters is unable to promote the mixing and mass transfer required. Nagata (1975) has reviewed mixing characteristics of highly viscous non-Newtonian solutions and the degree of mixing obtained with a variety of impeller designs. To achieve adequate mixing, irrespective of impeller design, a considerable power input is required, resulting in high operating costs. Polysaccharides are recovered from the fermentation broth by precipitation with a solvent, methanol is used extensively. Recovery of the solvent is essential for process economy. Many microbial polysaccharides are used in the food industry, requiring a high grade raw material and a pure product. Fungal polysaccharides, however, are predominantly used in the plastics, paints and related industries. This allows for reduction in operating costs and may be one of the reasons why the future market for these materials looks secure albeit not of the same size as the market for dextran and xanthan.

A number of surveys recently published serves to illustrate the current commercial interest in the production of microbial polysaccharides. Aspects of mixing and mass transfer, formation and structure and industrial production of polysaccharides have all been highlighted (Margaritis & Zajic, 1978; Slodki & Camus, 1978; Sutherland & Ellwood, 1979; Lawson & Sutherland, 1978).

WASTE MATERIALS

There are two aspects of waste materials which directly concern the fermentation industry. Firstly, the industry faces the problem of disposing of its own waste. This is not a 'civil engineering' problem, but a search for profitable ways of disposing the waste. More emphasis is currently put on the second aspect, however, namely the utilisation of waste material (or by-product) as a raw material for fermentation processes. Many waste products contain significant amounts of fermentable chemicals. Potentially this is a cheap raw material which will find more favour as the cost of high grade raw materials continue to increase.

Waste as raw material

Hitherto, wastes as raw material have been limited to processes with low profit margins. At the more sophisticated end of the industry the unit value of the product is high enough to cover high operating costs. Wastes have found application in the production of SCP and industrial organics, e.g. industrial alcohol and citric acid. At list of wastes potentially available as a fermentation raw-material is given in Table 2. The list, although long and varied, is by no means exhaustive. As indicated, a number of the wastes are being explored for SCP production, some industrial plants are already in operation.

A number of factors are taken into account when the 'suitability' of a waste is considered for SCP. The cost of obtaining approval for a microorganism as food or feed is prohibitive. Few additions can be expected to the microorganisms given in Table 3. It is important therefore, that the biological availability of a waste is acceptable to the 'SCP microbes'.

Wastes may also carry a number of undesirables, i.e. pathogenic or toxigenic organisms, trace metals, etc. This may well require substantial pretreatment. Location of the waste is important. As most wastes are employed in processes where the unit value of the product is low, large transportation costs cannot be met. Thus, the fermentation tends to be

Table 2. Wastes potentially available for SCP production

*	Sulphite pulp liquor
*	Wood waste hydrolysate
*	Bagasse, molasses
	Corn stover, cobs
	Oat husks, wheat bran, rice bran
	Spent rice bran
	Cereal straw (wheat, rice, grass-seed, millet)
	Spent hops
*	Distillers' slops
*	Whey (whole, ultrafiltrate)
	Shells (cocoa, coconut)
*	Pulp (tomato, coffee, banana, pineapple, citrus, olive)
*	Waste waters (olive, palm-oil, potato, date, citrus, cassava, rubber, fruit canning, vegetable canning, freezing, drying)
	Wash waters (dairy, cannery, confectionery, bakery, soft drinks, sizing, malting, corn-steep, etc.)
*	Lactate wastes (sauerkraut, other pickles)
*	Bakers' and confectioners' waste
	Leaf protein supernatant
	Paper, cellulose fines
	Urban refuse
	Feedlot residue
*	Pig manure
	Chicken manure

* Wastes evaluated for SCP production

Table 3. Status of microorganisms for SCP

(a)	*Saccharomyces*: food, accepted if grown on sugars (including whey) or ethanol (even if petrochemical based); feed, general acceptance.
(b)	*Candida*, *Hansenula*: wide acceptance as feed; acceptance for food not precluded.
(c)	*Paecilomyces*: accepted as feed in Finland, not precluded elsewhere.
(d)	*Pseudomonad* ('Pruteen'): approved marketing as feed was expected in 1978.
(e)	*Fusarium*: awaiting FAO approval as feed.
(f)	*Aspergillus*: test evaluation completed.
(g)	*Kluyveromyces*: feed, accepted in France.

relatively small scale, using locally produced waste materials. If agricultural wastes are used, seasonability will be important.

A notable contribution to biotechnology, in terms of utilisation of waste, process technology and PR has been provided by the Finnish 'Pekilo' process, producing SCP from sulphite liquor using *Paecilomyces*. It is currently limited to Finland, but tests have been carried out to establish plants in other parts of the world. There is a wealth of information available on the process. Data on the process performance, feed trials and process technology have been published (Romantechuk, 1975; Romantschuk & Lehtomaki, 1978). This is very refreshing and hopefully a trend which will be taken up by more members of the fermentation industry. Another well known process is the Swedish/Swiss 'Symba' process in which potato starch is utilised for SCP production. Two steps are required, in the first the dilute starch is hydrolysed by an amylolytic microorganism *(Endomycopsis)*. The product is then utilised by a conventional SCP organism.

Whey has received much attention as a potential SCP source. Unfortunately the 'conventional' SCP yeasts *Saccharomyces* and *Candida* do not utilise lactose. It has been claimed that *Paecilomyces* will do this, but more promising results have been obtained with *Kluyveromyces fragilis*. This is used in a French process currently producing lactose based SCP used as a calf milk replacement. The whey protein is firstly recovered as a higher grade protein product. The 'clear' lactose liquor is then used in the fermentation.

Production of industrial alcohol is an alternative to SCP from fermentantable wastes. With the 'energy crisis' interest has focused on industrial alcohol as an energy source and the process technology has recently been reviewed (Bu'Lock, 1979). Economic ethanol production requires a carbon source with a high proportion of hexose sugars. This limits the number of utilisable wastes to the various cellulosic wastes produced in agriculture and forestry. Activities in this field are described in a separate chapter.

In industrial alcohol, large scale continuous culture has found the outlet so long promised by SCP. To maximise the productivity, the fermentation must be run at a high cell density, obtained by cell retention and/or cell recycle. An interesting development is the vacuferm process, in which increased alcohol production is obtained by operating under reduced pressure (Ramalingham & Finn, 1977). Using both cell recycle and reduced pressure, Cysewski & Wilke (1977) increased the productivity to 82 mg/l/hr from 7 mg/l/hr in a conventional continuous process. Compared to the traditional batch fermentation, it is claimed that the continuous vaccuum process for production of industrial alcohol is economically more attractive (Cysewski & Wilke, 1978). Compressing the CO_2 to atmospheric pressure and recovery of the alcohol is energy demanding. It has been pointed out that there are still physiological and engineering problems associated with the process (Finn & Fiechter, 1979).

Continuous production of industrial alcohol using immobilised cells is feasible on a lab-scale (Marituna & Sinclair, 1978; Griffiths & Compere, 1976). Production of CO_2 causes problems if a packed column reactor is used for the alcohol production. The cell particles are packed in a column with the substrate running through and unless there is an escape route for the CO_2 the reactor will become blocked. Mavituna & Sinclair (1978) dispersed the cell particles in a stirred fermenter. The particles were retained in the system by using an internal cell feedback device. Potentially this is a better solution than the packed column, but with the present advances in 'conventional' alcohol production application of immobilised cells seems unlikely to reach industrial scale with the present technology.

Waste disposal

The present system of dumping residual biomass will soon be unacceptable and alternative ways of waste disposal must be found. Utilisation of the biomass for animal feedstuff has been tried to a limited extent in the pharmaceutical industry. The problems involved are obtaining approval for the microorganism as a feed, obtaining a defined microbial culture and extensive washing of the biomass to remove unwanted metabolites. Clearly, this is costly, defeating the object. The biomass may be used as a source of energy, in direct combustion or pyrolysis. This may require complete or partial drying. A third alternative is to use the biomass (and the spent process liquor) as a fertiliser. The solid waste has a high nitrogen content and considerable potential as a slow release fertiliser. Possible effects of unwanted metabolites must be considered, however.

PROCESS TECHNOLOGY

The decision to develop large scale SCP plants was very good for biotechnology, undoubtedly helping to lift it to its present standing. It is ironic that the decision may have been the wrong one, at least for the companies concerned. The majority of the proposed plants are now idle or were never commissioned. The benefit to biotechnology in general was in application of continuous culture, scale-up and oxygen transfer.

Continuous culture

The fermentation industry has shown great reluctance to employ continuous culture. There are a number of reasons for this, the risk of back-mutation, sterility problems and the cost of converting a process from batch to continuous operation. To change a profitable batch process may be questionable. In processes where the profit margin is low and the productivity in continuous exceeds that of batch culture, the advantages of continuous operation outweigh the disadvantages. In the large scale production of SCP, therefore, continuous culture is employed. To optimise the fermentation, process control is essential. Thus application of continuous culture also led to developments in on-line analysis. The use of enzymes as analytical agents received particular attention as described elsewhere in this chapter.

Scale-up and oxygen transfer

The problems produced by the increase in scale required by the SCP processes have been considerable. This is illustrated by the development of the 'Pekilo' process. In scaling up to a fermenter volume of $360m^3$ both the design and construction practice had to be revised (Romantschuk & Lehtomaki, 1978). The problem of oxygen transfer was highlighted by the increase in scale. To achieve high rates of oxygen transfer, without escalating the operating costs, trends in fermenter design show a move away from the stirred tank fermenter (Kristiansen, 1978). A number of new fermenter configurations, both research and production scale have been introduced. The emphasis is on low investment and operating costs without a reduction in the oxygen transfer capacity. Although most of the 'new' fermenters do not employ mechanical agitators, it must be stressed that the stirred tank is still the preferred design. It has been demonstrated, however, that for a specific process, a different configuration may represent a better choice. Tailoring a fermenter to a process may be dangerous, and only a stirred tank may be sufficiently versatile to meet the new demands should changes in the process be required.

CONCLUSIONS

Fungal biotechnology is still, as in the past, orientated towards the food industry. The fermentation industry has succeeded, however, in marrying new and old technology and are now producing the traditional product on a larger scale and much more efficiently. As the fermentation products have found wider application, fungal technology has steadily moved into new areas and this probably represents the greatest challenge to the biotechnologists. However, fungal biotechnology has advanced to a high level of technology and the future is met from a position of strength.

REFERENCES

AUNSTRUP, K. (1977). Production of industrial enzymes. *Biotechnology and fungal differentiation* (Eds. J. Meyrath & J. D. Bu'Lock) pp. 151-171. London, UK: Academic Press.

BARBESGAARD, P. (1977). Industrial enzymes produced by members of the genus *Aspergillus*. *Genetics and Physiology of Aspergillus* (Eds. J.E. Smith & J.A. Pateman), pp 391-404. London, UK: Academic Press.

BARKER, S.A. & SOMERS, P.J. (1978). Enzyme electrodes and enzyme based sensors. *Topics in enzyme and fermentation technology* 2 (Ed. A. Wiseman), pp. 120-151. Chichester, UK: Ellis Horwood.

BLAIN, J.A. (1975). Industrial enzyme production. *The Filamentous Fungi, Vol. 1* (Eds. J.E. Smith & D.R. Berry), pp. 193-211. London, UK: Arnold.

BOLASKI, W. & GALLANTIN, J.C. (19762). Enzyme conversion of cellulosic fibres. *US Pat. 3041246.*

BU'LOCK, J.D. (1979). Industrial Alcohol. *Microbial Technology: Current state, future prospects (29th Symposium of the Society for General Microbiology)* (Eds. A.T.Bull, D.C. Ellwood & C. Ratledge) pp 309-325. Cambridge, UK: Cambridge University Press.

BU'LOCK, J.D., DETROY, R.W., HOSTÁLEK, Z. & MUNIM-al-SHAKARI, A. (1974). Regulation of secondary biosynthesis in *Gibberella fujikorai*. *Transactions of the British Mycological Society* 62, 377-389.

CATLEY, B.J. & KELLY, P.J. (1975). Metabolism of trehalose and pullulan during the growth cycle of *Aureobasidium pullulans*. *Biochemical Society Transactions* 3, 1079.

CLARK, D.S. (1962). Submerged citric acid fermentation of ferrocyanide treated beet molasses: morphology of pellets of *A. niger*. *Canadian Journal of Microbiology* 6, 133-141.

CLARK, D.S. & LENTZ, C.P. (1963). Submerged citric acid fermentation of beet molasses in tank type fermenters. *Biotechnology and Bioengineering* 5, 193-199.

CYSEWSKI, G.R. & WILKE, C.R. (1977). Rapid ethanol fermentation using vaccuum and cell reycles. *Biotechnology and Bioengineering* 19, 1125-1143.

CYSEWSKI, G.R. & WILKE, C.R. (1978). Process design and economic studies of the alternative fermentation methods for the production of ethanol. *Biotechnology and Bioengineering* 20, 1421-1444.

ENFORS, S.O. & MOLIN, N. (1979). Enzyme electrodes for fermentation control. *Process Biochemistry* 13 (2), 9-11, 24.

FINN, R.K. & FIECHTER, A. (1979). Physiology and reactor design. *Microbial Technology: Current state, future prospects (29th Symposium of the Society for General Microbiology)* (Eds. A.T. Bull, D.C. Ellwood & C. Ratledge) pp. 83-105. Cambridge, U.K.: Cambridge University Press.

GRIFFITHS, W.L. & COMPERE, A.L. (1976). A method for coating fermentation tower packing so as to facilitate microorganism attachment. *Developments in Industrial Microbiology* 17, 241-246.

KRISTIANSEN, B. & SINCLAIR, C.G. (1979). Production of citric acid in continuous culture. *Biotechnology & Bioengineering* 21, 297-315.

LAWSON, C.J. & SUTHERLAND, I.W. (1978). Polysaccharides. *Primary products of metabolism (Economic Microbiology 2)* (Ed. A.H. Rose) pp. 327-392, London, U.K.: Academic Press.

LOCKWOOD, L.B. (1975). Organic production. *The Filamentous Fungi Vol. 1,* (Eds. J.E. Smith & D.R. Berry) pp. 140-157. London, U.K.: Arnold.

MARGARITIS, A. & ZAJIC, J.E. (1978). Mixing, mass transfer and scale-up in polysaccharide fermentations. *Biotechnology and Bioengineering* 20, 939-1001.

MAVITUNA, F. & SINCLAIR, C.G. (1978). The efficiency of gel-entrapped microorganisms as biocatalysts. *Paper presented at First European Congress of Biotechnology, Interlaken 1978.*

MESSING, R.A. (1975). *Immobilised enzymes for industrial reactors*.New York, U.S.A.: Academic Press.

MIALL, L.M. (1978). Organic acids. *Primary products of metabolism (Economic Microbiology 2)* (Ed. A. H. Rose), pp 47-119. London, U.K.: Academic Press.

MOSS, M.O. (1977). Enzymatic alterations of Penicillins and Cephalosporins. *Topics in enzyme and fermentation Technology* 1, (Ed. A. Wiseman), pp 111-131. Chichester, U.K.: Ellis Horwood.

NAGATA, S. (1974). *Mixing, principles and application.* New York, U.S.A.: Wiley.

OLSON, A.C. & COONEY, C.L. (1974). *Immobilised enzymes in food and microbial processes.* New York, U.S.A.: Plenum Press.

PERIMAN, D. (1975). Influence of penicillin fermentation technology on processes for production of other antibiotics. *Process Biochemistry* 10 (5), 23-27, 32.

RAMALINGHAM, A. & FINN, R.K. (1977). The vacuferm process: a new approach to fermentation alcohol. *Biotechnology and Bioengineering* 19, 583-589.

ROMANTSCHUK, H. (1975). Pekilo process. Protein from spent sulphite liquor. *Single cell protein* 2 (Eds. S.R. Tannenbaum & D.I.C. Wang), pp 344-356. Cambridge, Mass., U.S.A.: MIT Press.

ROMANTSCHUK, H. & LEHTOMAKI, M. (1978). Operational aspects of first full scale Pekilo SCP mill application. *Process Biochemistry* 13 (3), pp 16-17, 29.

RYDER, E.N. & STRONG, A. (1967). G.B. Patent 1085 901.

SCHIERHOLT, J. (1977). Fermentation processes for the production of citric acid. *Process Biochemistry* 12 (9), pp 20-21.

SLODKI, M.E. & CADMUS, M.C. (1978). Production of microbial polysaccharides. *Advances in Applied Microbiology* 23, pp 19-24.

STERNBERG, A. (1974). Production of cellulase by *Trichoderma. Biotechnology and Bioengineering Symposium No. 6*, pp 35-53. New York, U.S.A.: Wiley.

SUTHERLAND, I.W. & ELLWOOD, D.C. (1979). Microbial exopolysaccharides. *Microbial Technology: current state, future prospects (29th Symposium of the Society for General Microbiology)* (Eds. A.T. Bull, D.C. Ellwood & C. Ratledge), pp 107-150. Cambridge, U.K.: Cambridge University Press.

UNDERKOFLER, L.A. (1969). Development of a commercial enzyme process: glucoamylase. *Cellulases and their applications* (American Chemical Society Publication) (Ed. R.F. Gould), pp 343-358.
YUEN, S. (1974). Pullulan and its application. *Process Bio-Chemistry* 9 (9), 7-9, 22.
ZAJIC, J.E. & LeDUY, A. (1976). Pullulan. *Encyclopedia of Polymer Science*. New York, U.S.A.: Wiley.

FUELS FROM BIOMASS (FERMENTATION) - U.S.A.

T. Jeffries[1], B.S. Montenecourt[2]
and D.E. Eveleigh[2]

[1]*United States Department of Agriculture, Forest Products Laboratory, P.O. Box 5130, Madison, WI 53705*

[2]*Department of Biochemistry and Microbiology, Rutgers - The State University, New Brunswick, NJ 08903*

INTRODUCTION

Biotechnology in the United States has advanced broadly in such divergent areas as immobilization of whole cells and enzymes, application of computers to fermenter operation, and bio-separation technology (Wang, Lih & Hatch, 1977). One topic, the production of fuels from biomass, has received considerable emphasis as a result of the dwindling supplies of petroleum and the need to develop alternate renewable energy resources. Biomass such as forest, agricultural and animal residues, can be converted by physiochemical or fermentation processes to clean fuels, petrochemical substitutes or other energy intensive products such as ammonia. This article addresses the fermentation technologies being developed for these purposes through the Fuels from Biomass Program, U.S. Department of Energy (DOE).

U.S. ENERGY SUPPLY AND DEMAND

In order to evaluate the prospects for biomass as an energy resource in the U.S., it is necessary to understand the pattern of energy supply and demand. Fig. 1 illustrates the flow of energy in the U.S. for 1976 (Ramsey, 1977). In this graph, the widths of the lines are proportional to the quantity of the energy in each sector. The absolute amounts of energy in each sector are expressed as quads (1 quad = 1×10^{15} British Thermal Units). The energy flow is divided into supply, conversion and consumption sectors. For the latter two sectors, energy is shown as either rejected or useful. By this criterion, the overall efficiency of energy utilization is about 47%. In 1977, the latest year for which com-

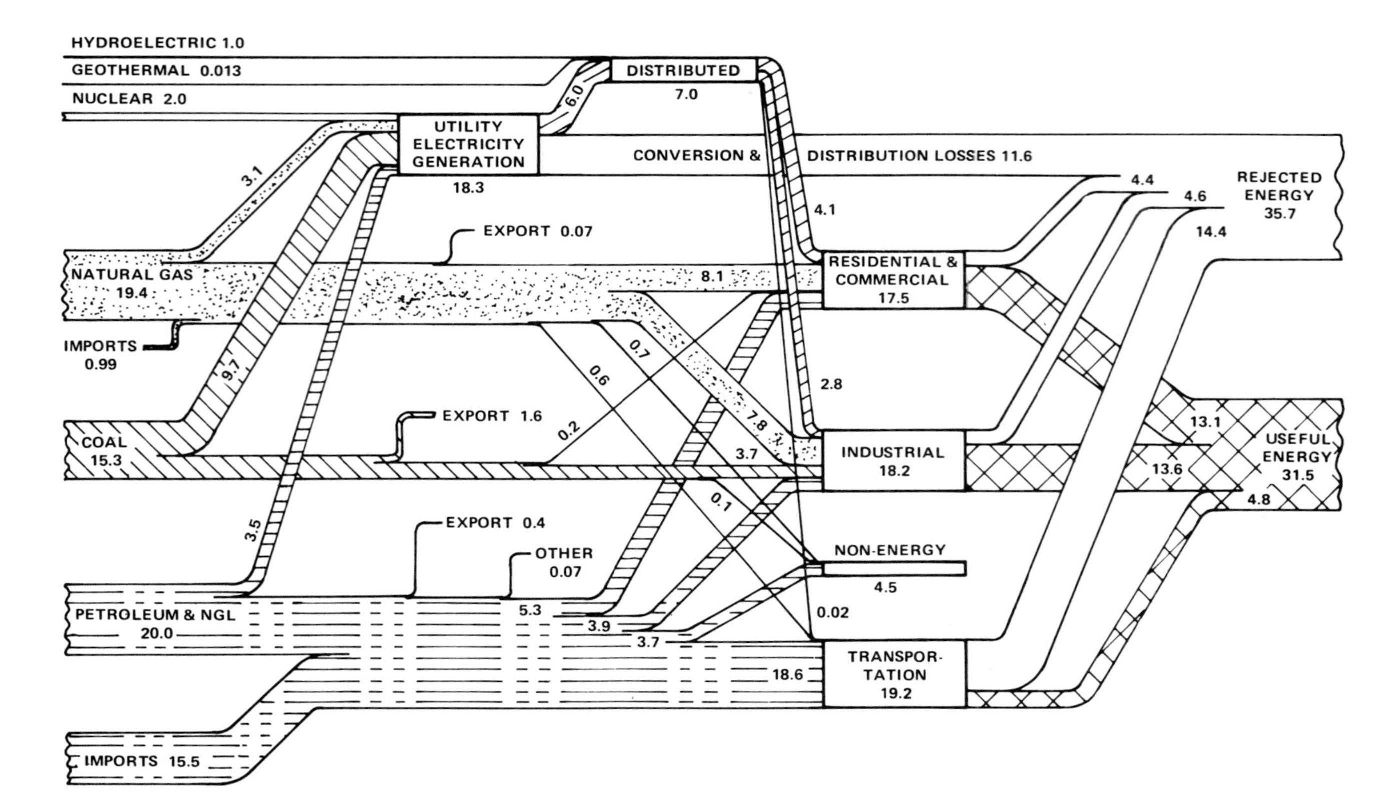

Fig. 1. U.S. Energy Flow 1976. (Primary resource consumption 72.1 quads)

plete figures are available, the domestic energy consumption was 76.2 quads. Of this total, 48% was supplied by petroleum. The percentage of energy consumed by the transportation sector was 26% in 1977 and the first 10 months of 1978 (DOE, 1979). However, unlike the other major energy consuming sectors, transportation energy demands are almost entirely dependent upon petroleum. That is to say, they are relatively inflexible (Werth, 1976). By comparison, utility electrical generation, residential and commercial heating and industrial users of energy derive only 20 or 30% of their energy from petroleum, and they can be more readily adapted to natural gas, coal and solar energy resources. Petrochemical and transportation demands draw heavily upon liquid fuels. Over the last three years, approximately 49% of the petroleum consumed in the U.S. has been imported, and about 54% of the total was used for domestic transportation. It is on this basis of the need for liquid transportation fuels that DOE has emphasised the production of ethanol in their Fuels from Biomass Program.

The Potential Contribution of Biomass to the U.S. Energy Budget

The present total annual biomass production in the U.S. has been estimated to be equivalent to about 27 quads (Benemann, 1978). Of this total, about 1 billion tons of dry organic matter is generated as waste (Anderson, 1977). By presently available conversion methods, this material could yield about 8 to 10 quads of liquid or gaseous fuels (Anderson, 1972). The amount of waste material that is or could be economically available varies with the collection and conversion technologies employed (Gregor & Jeffries, 1979), but there is a general consensus that economically available biomass from agricultural, municipal and silvicultural resources could amount to around 7 quads, or about 7% of the projected energy demand by the year 2000 A.D. (Anon, 1979) (Fig. 2). Although it is not generally included in assessments of the U.S. energy budget, organic renewable resources presently contribute about 1.8 quads to the U.S. energy supply (Tillman, 1977). This value compares favourably with the 1.8 quads of electricity presently supplied by nuclear generating plants in the U.S.

As fossil fuel resources are depleted or become more constrained by environmental considerations, conversion of organic residues to liquid fuels will become more economically attractive. More marginal lands could be brought into use specifically for biofuel production, and it is conceivable that eventually the U.S. could derive 15 to 20 quads of fuels annually from wastes and other terrestrial biomass resources (Del Gobbo, 1978).

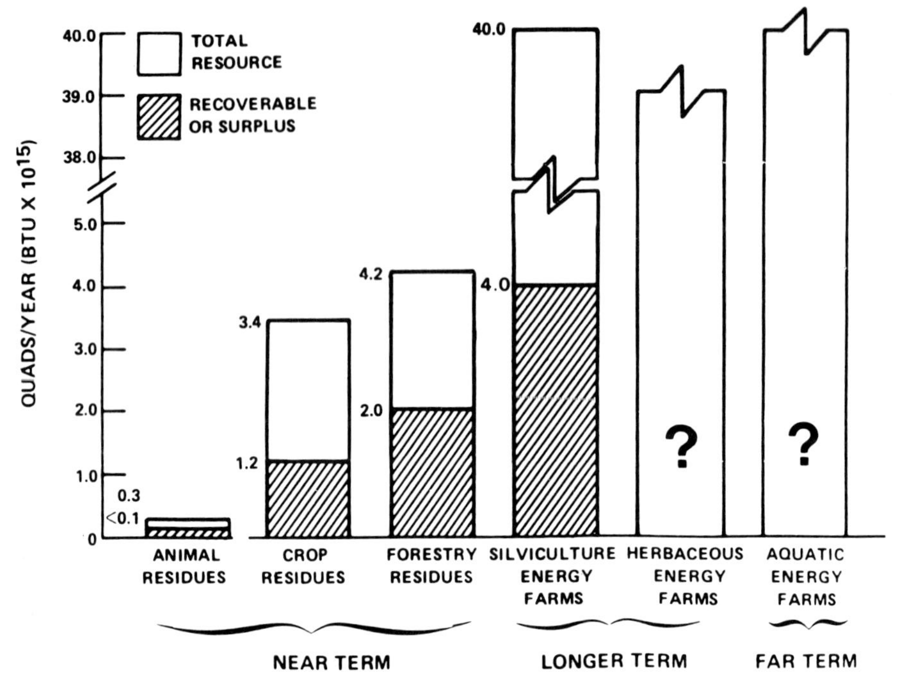

Fig. 2. U.S. Biomass reserve availability

Types of Biomass

The production of alcohol by fermentation of sugars and starch is an ancient art. Though of current marginal economic payoff, large scale fermentative production of ethanol in the U.S. from maize has been proposed (Scheller, 1976). The U.S. Department of Agriculture has guaranteed four $15 million dollar loans to aid, in the setting up of demonstration plants (U.S. Federal Register, 1978). However, as the global food supply cannot meet the world's needs, alternate sources of biomass should be considered as an energy resource. The most readily available alternate fermentation substrate is cellulose; for instance in the U.S. from hardwoods, or from maize and sweet sorghum stover. The federal agencies have programs to assess the potential application of these substrates. The National Science Foundation has developed a long range program on the utilization of lignocellulose as a source of chemicals (Gainer, 1976; National Science Foundation, 1977). In contrast, the DOE program has focussed on a shorter term program with cellulose as the major substrate for the production of "Fuels from Biomass".

FUELS FROM BIOMASS (FERMENTATION)

The DOE Fuels from Biomass (FFB) program is wide ranging and covers biomass production (via agriculture, silviculture, or aquatic systems), photoelectrolysis, and thermochemical and biochemical conversion to produce fuels and petrochemical substitutes. For review see (Anon, 1978; Schuster, 1978; U.S. Dept. Energy, 1978). Funding of the FFB program has increased considerably over the last few years to $28 million dollars for 1978 (Fig. 3).

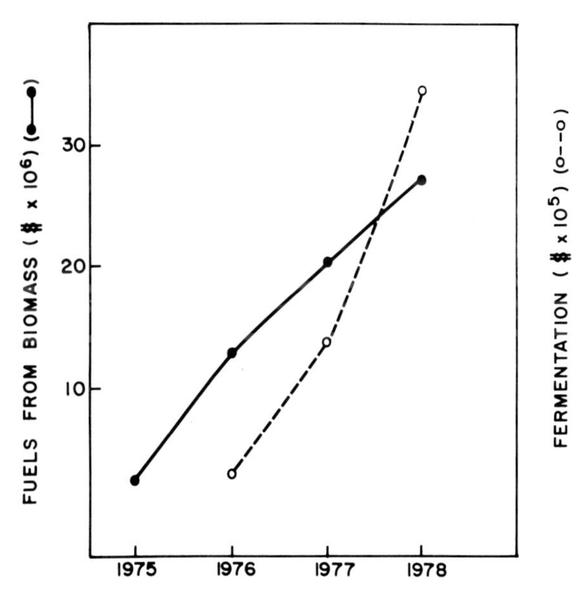

Fig. 3. D.O.E. trend in funding for fermentation

The "Biochemical Conversion" facet is broadly divided into "Anaerobic Digestion" for production of gaseous fuels, and "Fermentation" for preparation of liquid fuels (Fig. 4). The $3 million 1978 fermentation budget was distributed amongst Universities (50%), contractors (40%), small business contractors (9.9%) and other federal agencies (0.1%). These proportions will probably change with the inception of the new DOE laboratories - the Solar Energy Research Institute (SERI), Golden Colorado.

The current FFB-Fermentation projects and contractors are outlined in Table 1 and an integrated overview of the processes and their potential products in Fig. 5. They all use cellulosic biomass as a substrate. As noted above this is the most widely available abundant renewable resource and is economically attractive when full credit is taken for

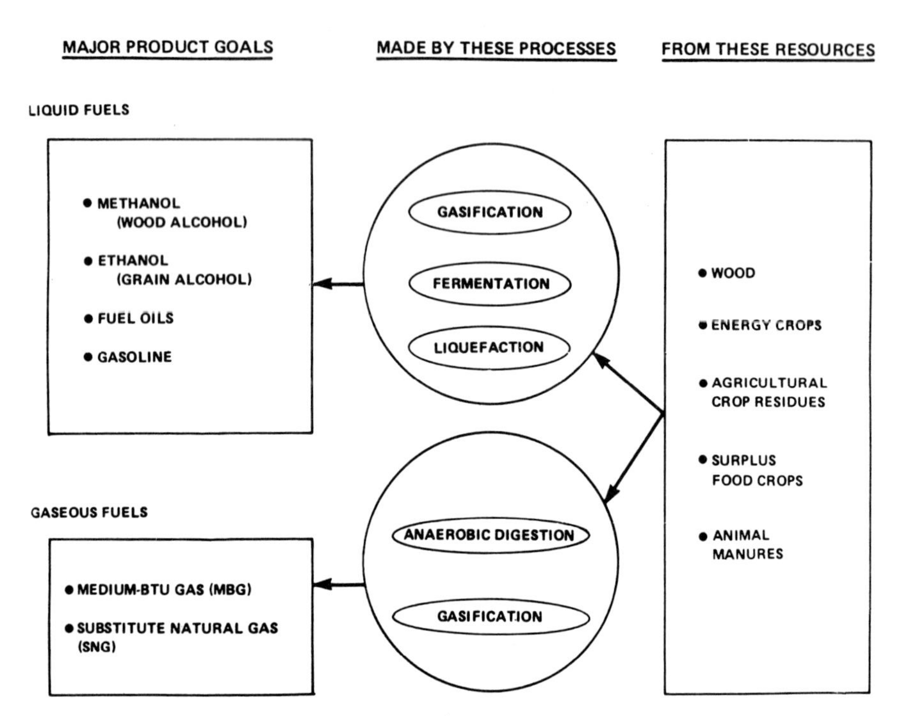

Fig. 4. D.O.E. Program Overview

its lignin and hemicellulose components and for the stillage derived from its fermentation.

The utilization of biomass in the FFB program is subdivided into four research areas (Fig. 5).

Extraction and use of hemicellulose
Separation of lignin
Cellulose - indirect conversion
Cellulose - direct conversion

Extraction and use of hemicellulose

Efficient extraction of hemicellulose from corn stover is accomplished with sulphuric acid (1.9%) at 121°C (PU)*.

*(PU) Contractor abbreviation, see Table 1.

Xylan, the principal product, can be directly fermented to ethanol *(Fusarium oxysporum)* or to ethanol and acetone *(Bacillus macerans* (B). Xylose is readily obtained by enzymatic degradation of xylan by a variety of microbial xylanases (B,N,UPG). Xylose can be fermented directly by *Clostridium thermosaccharolyticum* to ethanol and organic acids (M) and by *Aeromonas hydrophila* to 2-3 butane-diol in 60% yield (Pu). Xylitol, a tooth decay preventive, can be produced by chemical reduction of xylose.

Separation of Lignin

The processes under study include: (1) Solvation of cellulose; (ii) Solvation of lignin; (3) Physico-chemical separation of cellulose and lignin; and (iv) double roll milling.

(i) Solvation of cellulose (Pu). A major advance is the solvation of cellulose in 70% H_2SO_4 from lignocellulose. The extracted cellulose is reprecipitated in methanol and is amenable to enzymic or acid hydrolysis. The insoluble lignin is used as a fuel. This process could be the key economic step to the use of cellulosic biomass.

(ii) Solvation of lignin (UPG). Effective solvation of lignin in butanol is an attractive alternate process especially as one of the later products from a clostridial cellulose fermentation, is butanol. The butanol/lignin mixture can be used as a bunker fuel.

(iii) Physico-chemical separation of cellulose and lignin - "High pressure steam explosion". Lignocellulose is super heated in steam under pressure, then "exploded" by rapidly dropping the pressure back to atmospheric. This results in "liquefaction" and dissociation of the lignocellulose complex. The cellulose product is amenable to enzymatic saccharification and is also more digestible in the ruminant. Companies exploring this technology include:

Iotech Corporation Limited, Kanata, Ontario, Canada
Stake Technology Limited, Ottawa, Canada

The General Electric Schenectady (GE) group is developing this process in combination with the use of sulphur dioxide to obtain the greatest yield of hydrolyzable products.

(iv) Double roll milling (N). Direct physical abrasion by ball milling or by double roll milling considerably enhances the susceptibility of cellulose to enzymatic attack, but the expense of these processes is relatively high.

Table 1. Fuels from Biomass (Fermentation), US Department of Energy (1978)

Approach	Emphasis	System	Institution	Investigators
Pretreatment	I. Hemicellulose Extraction	Dilute Acid H_2SO_4	Purdue Univ.(PU)	M.C. Flickinger M. Ladisch G.T. Tsao
	IIa. Separation of lignin from cellulose	(1) 70% acid H_2SO_4	Purdue Univ	
		(2) n-Butanol extraction	Univ. of Pennsylvania/ General Electric (king of Prussia) (UPG)	J. Forro A.E. Humphrey E. Nolan E.K. Pye
	IIb. Physical pretreatment	(1) Steam explosion + Sulphur dio-xide	General Electric, Schenectady (GE)	W.D. Bellamy R. Brooks T.-M. Su
		(2) Two roll milling	U.S. Army Natick Labs., Natick, Mass (N)	T. Tassanari
Indirect Fermentation	I. Acid saccharification of cellulose	(1) Dil.H_2SO_4/ 235^oC	Darmouth Univ. (D)	A.O. Converse H.E. Grenthlein
		(2) Dil.H_2SO_4/ 125^oC	Purdue Univ.	

contd.

II. Enzymatic saccharification of cellulose	(1) Mutant screening; saccharification; cellulase characterisation; pilot plant enzyme production. *Trichoderma*	Natick (N)	F. Bisset M. Mandels E.T. Reese L. Spano D. Sternberg
	(2) Saccharification; bench and continuous fermentations (high density and cell recycle) for cellulase production; vacuum fermentation (ethanol production) *Trichoderma/ Saccharomyces*	Univ. of California Berkeley (B)	H. Blanch S. Rosenberg C.R. Wilke R.D. Yang
	(3) Mutation of microbes to increase cellulase yields *Trichoderma*	Rutgers Univ.	D.E. Eveleigh B.S. Montenecourt
	(4) Cellulase *Thermoactinomyces*	Univ. of Pensylvania/ General Electric	
	(5) Cellulase *Clostridium thermocellum*	Massachusetts Institute of Technology (M)	

contd.

Approach	Emphasis	System	Institution	Investigators
	III. Hemicellulose utilization	(1) 2-3 Butanediol from *Aeromonas & Klebsiella*	Purdue Univ.	
		(2) Ethanol from either *Fusarium* or *Bacillus macerans*	Univ. of California Berkeley	
		(3) Xylitol (Xylanase + Reduction)	Univ. of Pennsylvania/ General Electric	
Direct Fermentation	I. Clostridia	(1) Ethanol, acetic acid, acetone, n-Butanol *C. acetobutylicum* *C. thermoaceticum* *C. thermocellum*	(a) Massachusetts Inst. of Technology	C.L. Cooney A. Demain R. Gomez A. Sinskey D.I.C. Wang
		C. *thermocellum*	(b) G.E., Schenectady	
		Simultaneous conversion of cellulose by *Thermoactinomyces* cellulase plus; (a) *C. acetobutylicum* (b) *C. thermocellum*	(c) Univ. of Pennsylvania/ General Electric	
	II.	(2) Acrylic acid *C. propionicum*	M.I.T.	

contd.

	III. Algal biomass	Mixed anaerobes/ +Kolbe electrolysis- alphatic hydrocarbons	Dynatech. Cambridge Massachusetts	J.E. Sanderson D. Wise
Low Scale Technology	Program management; Low technology fermentation		Rensselaer Polytechnic Institute	H.R. Bungay

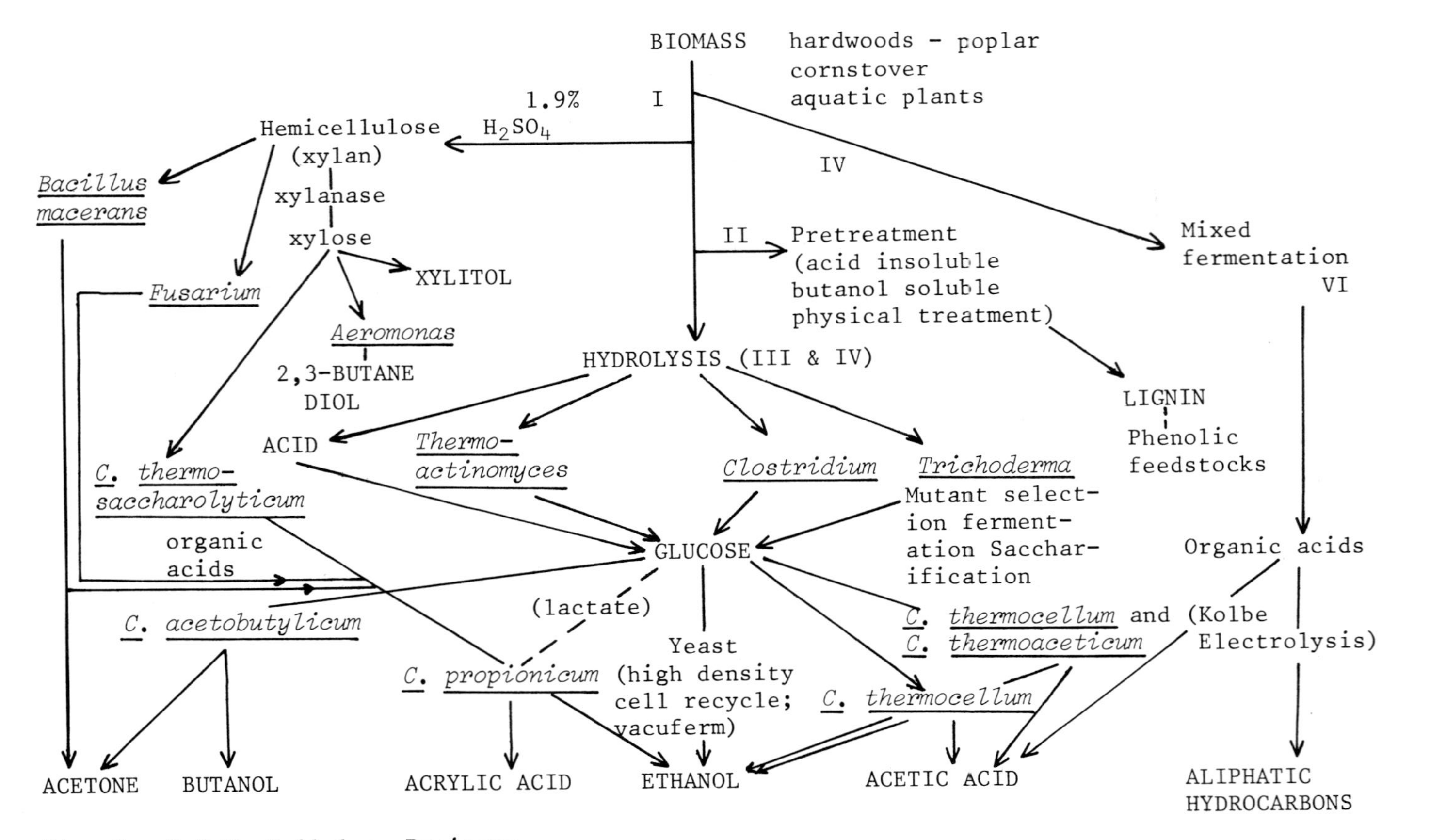

Fig. 5. D.O.E. Cellulose Projects

Cellulose - Indirect Conversion

In this scheme cellulose is first converted to glucose and the latter subsequently fermented to a fuel or chemical foodstock. Both acid and enymatic hydrolysis of cellulose for the production of glucose have been considered as each method has its attributes. Analogously, both approaches are used for glucose production in the starch industry. The effectiveness of acid hydrolysis of cellulose is being examined at Auburn, Dartmouth and Purdue Universities. Hydrolysis occurs in less than 1 min at temperatures of 230-240°C, but glucose yields are about 50% with considerable amounts of furfurals and other by-products (D). Optimization to increase the yields of glucose is under study and also processes to allow the use of higher cellulose slurry concentrations (initially 5%, a target of 20%). In the Purdue system, after solubilization of the cellulose from lignocellulose in 70% H_2SO_4, it is subsequently hydrolyzed in 0.5% H_2SO_4 at 125°C.

Enzymatic hydrolysis of cellulose has received considerable attention; 100% conversion of cellulose to glucose is possible but there are two major bottlenecks to this approach. First the substrate must be delignified in order to permit effective enzymatic hydrolysis (see above). Secondly, the cost of the enzyme must be drastically reduced. The latter is addressed in the following program.

A considerable understanding of the nature of the cellulase complex has been gained through long term studies at the U.S. Army Natick Laboratories. The mesophilic fungus *Trichoderma reesei*, was selected as one of the few producers of a cellulase that effectively hydrolyzes crystalline cellulose. This enzyme works effectively at 50°C. The optimal conditions for pilot plant production of *Trichoderma* cellulase have been elucidated and a range of methods for promoting optimal saccharification of raw cellulosics has been assessed at Natick. The development of high yielding cellulolytic mutants at Rutgers University and at Natick in combination with the previously mentioned optimized fermentation technology, has resulted in considerably more efficient production of cellulase and removed much of the problem of cost of cellulase. Current yields of 15 filter paper units (FPU)/ml (productivities at 100 FPU/l/h) have been attained in comparison to 0.5 FPU/ml by the wild strain when grown under shake flask conditions. Production of cellulase in continuous mode including cell recycle developed at the University of California, Berkeley, raises the possibility of even less expensive cellulase production. Recovery of cellulase by desorption of the enzyme from spent cellulose using urea, appears a reasonable cost effective process.

A cellulase from *Thermoactinomyces* that is stable at 60°C has been isolated at the University of Pennsylvania, and shows good promise.

Ethanol production by yeast from glucose is the final stage. At Berkeley, improvements in ethanol production have been achieved through use of both high density and cell recycle culture techniques, and also through vacuum fermentation for recovery of ethanol.

Cellulose - Direct Conversion

In this approach a single direct fermentation of cellulose is proposed as being the most economically attractive route. The products, ethanol and potential chemical feedstocks, are considered as useful energy sparing products. Clostridia are of key importance in this approach with the basic attributes of being anaerobic, thermophilic and with efficient production of a range of useful products; for instance *Clostridium thermoaceticum* produces acetic acid from glucose in 85% yield.

Clostridium thermocellum has the key role in this scheme as it is cellulolytic. Two approaches to its use include homoculture fermentation to yield ethanol and acetic acid or mixed culture fermentations to gain greater yields of a particular product; for example acetic acid by mixed culture of *C. thermocellum* and *C. thermoaceticum*. Both approaches appear useful. In pure culture *C. thermocellum* produces roughly equal amounts of ethanol (4 g/l) and acetic acid (M, GE, UPG). Ethanol resistant strains (*C. thermocellum* S-4, (M)), isolated through "adaptive selection", efficiently produce nearly twice the amount of ethanol (9 g/l). Screening is being continued as higher product concentrations are necessary on an economic basis (M, GE). *C. thermocellum* is also grown on cellulose in combination with *Thermoactinomyces* cellulase plus a vacuum recovery of ethanol in order to promote a more effective fermentation (UPG). Similarly, *C. acetobutylicum* in the presence of the cellulase, is grown on the products of hydrolysis of cellulose in the production of acetone and butanol (UPG). Mixed cultures of *C. thermocellum* and *C. thermoaceticum* have been shown to optimise acetic acid production (M). A dual culture approach for acrylic acid production envisages production of lactic acid by Lactobacilli and its conversion to acrylic acid by *C. propionicum*. Mutants of *C. propionicum* have been isolated that are blocked at the acrylic acid stage of their pathway from lactate to propionate. (M). This could be a most useful approach to producing a chemical feed stock as well over 500 million pounds of acrylates are used annually.

A unique mixed culture system to produce aliphatic hydrocarbons from aquatic cellulosic biomass (*Hydrilla*) is being explored by the Dynatech Corporation. This process entails production of organic acids in mixed culture in anaerobic packed bed fermenters, their extraction (liquid-liquid) and subsequent low voltage electrolytic oxidation (Kolbe electrolysis) to aliphatic hydrocarbons. The process gives reasonable yields and includes minor amounts of olefins, esters and alcohols as by-products. The projected cost of the product is $5.30/MMBtu when starting with a biomass raw material cost of $30/ton.

COMMUNICATION AND LOW SCALE TECHNOLOGY

Rensselaer Polytechnic Institute oversees the Fermentation Program and gives communication between the contractors the highest priority. Quarterly meetings allow discussion of new advances and also allow focus on problems which are often resolved by joint discussion. The scientific program is monitored by the use of outside consultants, while the economic outlook is reviewed by expert analysts (Batelle Laboratories, Columbus, Ohio; S.R.I. International, Melno Park, Ca.). A demonstration project is currently being designed based on the results of the overall program (Georgia Institute of Technology). Rensselaer Polytechnic also has a project on low investment technology. Concepts under study include use of non-sterile fermentations through partial pasteurisation, more effective sedimentation of yeast in a plate settler and maintenance of high cell concentrations in high dilution rate continuous cultures through use of a substrate feedback control (Bungay, 1978).

DISCUSSION

This brief review covers the DOE FFB-fermentation program which focusses on the production of liquid transportation fuels from biomass. This integrated approach with marked feedback between the university and industrial contractors (see Fuels from Biomass Newsletters), has resulted in considerable advances over a broad range. Ethanol has been stressed as a product, although its use as a fuel based on old technology is not overly optimistic (Park, Price & Salo, 1978). However, such analyses must be seen in perspective. First, they do not take into account the benefits to be derived from future technological advances (Gregor & Jeffries, 1979). Yet these advances are the *quid pro quo* of the Biomass Fermentation Program. For instance, a single fermentation to quantitatively convert cellulose directly to ethanol would produce considerable cost benefit. This is possible by preparation of cellulolytic yeast or zymomonad; an achievable goal through current genetic engineering techniques. With subsequent cell immobilization techniques this could prove a most economically attractive route. Secondly, gas and oil are highly subsidized for exploration and production and when these subsidies are taken into account the potential cost of biofuels appears quite favourable. A third and more complex economic factor is the costs associated with environmental damage incurred in the use of fossil fuels. Such cost factors, arising from pollutants such as acid rain, from oil spills or stripmine land damage, are high. They are insignificant in the use of biofuels.

In summary, biofuels are important in that they can provide up to 10% of theU.S. energy needs on a continuing basis. In a parochial vein, the development of biofuel has resulted in considerable advancement in biochemical engineering technology.

ACKNOWLEDGEMENTS

The authors wish to express their gratitude to Dr. W.J. Ramsey and The Energy and Resource Planning Division of Lawrence Livermore Laboratory for supplying the authors with Fig. 1, and for expert technical assistance in the assessment of U.S. Energy supply and demand.

REFERENCES

ANDERSON, L.L. (1972). Energy potential from organic wastes: a review of the quantities and sources. U.S. Department of the Interior, Bureau of Mines, Information Circular 8549. Washington, D.C.

ANDERSON, L.L. (1977). A wealth of waste; a shortage of energy. pp. 1-16. In *Fuels from Waste*, Ed. L.L. Anderson and D.A. Tillman, Academic Press, N.Y.

ANON (1978). Annual Review of Solar Energy - Program Review 1977, Solar Energy Research Institute. SERI/TR-64-066.

ANON (1979). Biomass potential in 2000 put at 7 quads. *Chemical and Engineering News*, February 12, 1979, p. 20.

BENEMANN, J.R. (1978). Biofuels: A Survey. Electric Power Research Institute, Palo Alto, Ca., U.S.A. Special Report June, 1978, ER-746-SR.

BUNGAY, H.R. (1978). Low technology fermentation. In *Second Annual Fuels from Biomass Symposium*, Ed. W.W. Schuster, pp. 609-611. Rensselaer Polytechnic Institute, Troy, N.Y.

DEL GOBBO, N. (1978). Fuels from biomass systems program - overview. In *Second Annual Symposium on Fuels from Biomass*, Ed. W.W. Schuster, pp. 7-24. Rensselaer Polytechnic Institute, Troy, N.Y.

DEPARTMENT OF ENERGY, U.S. Monthly Energy Review. January 1979. Energy Information Administration, Washington, D.C. NTIS UB/E/127-001.

GAINER, J.L. (1976). Enzyme technology and renewable resources. Proceedings Grantees-Users Conference, May 19-21, 1976. National Science Foundation (RANN) NSF/RA 760180.

GREGOR, H.P. & JEFFRIES, T.W. (1979). Ethanolic fuels from renewable resources in the solar age. *Annals of the New York Academy of Science* (In Press).

NATIONAL SCIENCE FOUNDATION (1977). Summary of Awards: Advanced Energy and Resources Research. *Biomass Technology* pp. 58-68 NSF/RA-770390.

PARK, W., PRICE G. & SALO, D. (1978). Biomass-based alcohol fuels: The near term potential use with gasoline. A report by the Mitre Corporation. U.S. Department of Energy Report HCP/T4101-03. UC-61.

RAMSAY, W.J. (1977). U.S. Energy Flow in 1976. Lawrence Liverpore Laboratory, Livermore, Ca.

SCHELLER, W.A. "Nebraska 2 Million Mile Gasohol Road Test Program", 6th Progress Report, University of Nebraska, Lincoln, NE, April 1976 - December 1976.

SCHUSTER, W.W. (Ed.) (1978). *The Second Annual Symposium on Fuels from Biomass*, June 20-22, 1978, Rensselaer Polytechnic Institute, Troy, N.Y.

TILLMAN, D.A. (1977). Unaccounted energy : The present contribution of renewable resources. pp. 23-54. In *Fuels and Energy from Renewable Resources*, Ed. D.A. Tillman, K.V. Sarkanen and L.L. Anderson. Academic Press, New York.

U.S. DEPARTMENT OF ENERGY FUELS FROM BIOMASS (1978). Program Summary, January 1978. DOE/ET-002/1/UC-61.

U.S. FEDERAL REGISTER (1978). 43 No. 133 July 11, United States Department of Agriculture: Commodity Credit Corporation.

WANG, D.I.C., LIH, M.M. & HATCH, R.T. (1978). Workshop on the Fundamentals of Biochemical Engineering. Final Report NSF Grant No. ENG-78-00726.

WERTH, G.C. (1976). Changing Energy Perspectives. Lawrence Livermore Laboratory, Livermore, Ca. UCRL-78153.

ADDENDUM

The most recent data are available from Government Reports (Contractors Quarterly Reports, National Technical Information Service, Springfield, Va) and brief updates are published in the Fuels from Biomass Newsletter, available from H.R. Bungay, Department of Chemical and Environmental Engineering, Rensselaer Polytechnic Institute, Troy, N.Y.

AN APPRAISAL OF ADVANCES IN BIOTECHNOLOGY IN CENTRAL AFRICA

J.A. Ekundayo

Department of Biological Sciences,
University of Lagos,
Lagos, Nigeria.

INTRODUCTION

In most developing countries, fermented food products constitute a substantial fraction of the diet. These products are usually obtained as a result of fermentation of different carbohydrate-rich raw materials, viz, stem and root tubers (e.g. *Manihot utilissima* Pohl, *Dioscorea* spp. L., *Ipomoea batatas* L. and Lam.) cereal grains (e.g. *Sorghum vulgare - S. bicolor* (Moench), *Zea mays, Pennisetum typhoideum*), and various fruits.

Most of the raw materials for the fermentation have low protein and vitamin contents. Consumption of such predominantly starchy materials give rise to malnutritional diseases such as kwashiokor and beri-beri.

Platt (1964) suggested that fermentation of carbohydrate-rich food materials could improve the nutritional value of the native South Africans - a process he described as "Biological Ennoblement".

To achieve this objective, international biotechnological programmes emphasize the possibilities of nutritional improvement through fermentation of low protein, carbohydrate-rich food crops and other fermentable products by means of microorganisms (especially bacteria and fungi including yeasts) or their enzymes.

In Central Africa, especially along the West African Coast and in some East African States, biotechnological studies involving the fermentative production, nutrient enrichment and detoxification of foods including beverages have become emphasized.

This article is intended to highlight the role of biological agencies in the production of indigenous non-alcoholic fermented foods and alcoholic beverages, and to bring into focus the biotechnological advances in the production of these fermented foods in Central Africa.

The fermentations described in the following sections are listed in Table 1. For convenience they are grouped as follows:

(a) Indigenous fermented alcoholic beverages

(b) Indigenous fermented non-alcoholic foods

(c) Wastes utilization.

The fermentation processes involved in the production of the various foods are carried out by a mixed flora including bacteria and fungi under controlled non-aseptic conditions; pure cultures have not been employed in any of the processes. The advances, if any, made in biotechnology in such processes have been aptly described as progress in microbial biotechnology, but in processes where fungi are implicated, their role will be emphasized to fit the article into the central theme of the Symposium.

ALCOHOLIC BEVERAGES

Palm wine and its derivatives

Palm wine is an alcoholic beverage produced from the sap of various palms. It is an important fermented beverage in West Africa and its production is developing into a major industry particularly in Nigeria and the Cameroons. Traditionally, it is obtained as fermentable sap of the raphia palms (*Raphia hookeri*, *Raphia vinifera*) and the palm tree (*Elaeis guineansis*) (Faparusi, 1966).

Palm juice is obtained by tapping the young flowering spathe of the respective palms. However, the method of tapping varies from one locality to another. The conventional method described by Bassir (1962) and Faparusi (1973) involves the removal of the leaf subtending an immature male inflorescence to obtain access to the inflorescence within the spaths. An incision is subsequently made near the apex of the inflorescence. The cut is covered by a piece of felt made of the fibrous leaf sheath fabric and a new cut is made daily until the juice begins to flow. The fresh palm juice which is referred to as palm wine is neutral in reaction. It is a sweet, clear, colourless syrup containing 10-12% sugars (Bassir, 1962).

One of the chief problems of the palm wine industry is

Table 1. Food fermentations in Central Africa

Name of product	Organisms involved	Substrate	Nature of product	Area of production
a) ALCOHOLIC BEVERAGES				
i. Palm wine	*Saccharomyces cerevisiae* *Candida tropicalis* *Candida utilis* *Schizosaccharomyces pombe* *Lactobacillus plantarum* *Lactobacillus brevis* *Leuconostoc mesenteroides* *Streptococcus* spp *Sarcina* spp *Bacillus* spp *Penicillium* spp *Aspergillus* spp	Fresh palm juice	Whitish liquid	Southern Nigeria, Cameroons
ii. Ogogoro (Native gin)	*Saccharomyces cerevisiae* *Candida* spp *Schizosaccharomyces pombe*	Fermenting palm juice	Colourless liquid	Southern Nigeria, Ghana, Cameroons
iii. Pito	*Rhizopus oryzae* *Aspergillus flavus* *Penicillium funiculosum* *Penicillium citrinum* *Botryodiploidia theobromae* *Geotrichum candidum* *Candida* spp	Sorghum, maize or a mixture of sorghum and maize	Dark brown liquid	Nigeria, Ghana

	Aspergillus versicolor *Penicillium simplicissimum* *Penicillium purpurogenum* *Lactobacillus* spp			
iv. Burukuto beer	*Saccharomyces* *Candida* *Hansenula anomala* *Kloeckera apiculata* *Mucor rouxii* *Aspergillus* spp *Penicillium* spp *Lactobacillus* spp *Streptococcus* spp *Acetobacter* spp	Sorghum plus gari	Creamy liquid	Nigeria, Republic of Benin, Northern Ghana, East Africa
b) NON-ALCOHOLIC FERMENTED FOODS				
i. Gari	*Corynebacterium manihot* *Geotrichum candidum* *Leuconostoc* *Lactobacillus* *Alcaligenes*	Cassava (*Manihot utilissima*)	White or yellow grains	West Africa
ii. Ogi	*Saccharomyces cerevisiae* *Candida mycoderma* *Lactobacillus plantarum* *Aerobacter cloacae* *Cephalosporium* spp	Maize, millet	White or brown cake	Nigeria, Ghana, Republic of Benin

contd. overleaf/

Table 1 contd.

Name of product	Organisms involved	Substrate	Nature of product	Area of production
ii. Ogi	*Fusarium* spp *Penicillium* spp *Aspergillus* spp			
iii. Soy ogi	As for Ogi (above)	Maize + soya beans	Brown powder	Nigeria
c) WASTES UTILIZATION				
i. Alcohol and vinegar	*Saccharomyces cerevisiae* *Schizosaccharomyces pombe* *Acetobacter* spp	Coffee pulp	Brown liquid	Nigeria, Cameroons, East Africa
ii. Vinegar	As for palm wine	Fermenting palm pulp	Clear liquid	Nigeria
iii. Single cell protein	*Geotrichum candidum* Local yeast strains	Cassava wastes	Microbial cell mass	Nigeria

microbial contamination of the palm juice as it oozes out of the palm tree. This contamination has made it difficult to collect the palm juice aseptically and this has placed a heavy limitation on the distribution of this product to distant places. It ferments rapidly with a drastic decrease in the sugar content; the sap turns milky white and sour within a period of 24 hours due to microbial growth and becomes unacceptable for drinking.

Analysis of palm wine microflora (Bassir, 1962; Faparusi, 1966, 1971, 1973; Faparusi & Bassir, 1972*a*,*b*; Okafor, 1972*a*,*b*, 1975*a*,*b*) has revealed the presence of a variety of microorganisms. They included *Saccharomyces cerevisiae*, *Schizosaccharomyces pombe*, *Candida utilis*, *Candida tropicalis*, *Micrococcus* spp, *Sarcina* spp, *Acetobacter* spp, *Streptococcus* spp, *Leuconostoc* spp, *Bacillus* spp, *Zymomonas* spp, *Serratia* spp, *Brevibacterium* spp, *Aerobacter (Klebsiella)* spp, and *Lactobacillus* spp. Occasionally *Penicillium* spp and *Aspergillu* spp were encountered. The yeasts and bacteria are known to originate from the indigenous flora of the palm trees and from equipment such as tapping knives, funnels and the collectin vessels (Okafor, 1972*b*; Faparusi, 1973).

During fermentation of palm wine, large quantities of carbon dioxide are produced. The production of carbon dioxide results in the development of high pressure in the containers which, if made of some weak material, may explode.

Conservative estimate of palm wine production in Nigeria stands at a daily production of 90 - 130 thousand litres. The wine is produced by individual tappers and pooled before being offered for sale. Since no suitable method of preservation has been evolved, much of the wine is either wasted if not consumed at the end of the day, or the left-over may be distilled for alcohol. To make palm wine available throughout the year, a successful preservation method giving a shelf life of at least six months is desirable since production is highest in the rainy season (March - October) and much reduced in the drier periods of the year (November - February).

A crude method of preservation involves the direct addition of the bark of a tree *Saccoglottis gabonensis* urban, (family Humiaceae), to fresh palm juice. The sap from the bark of the tree imparts an amber colour to the wine and it also gives the wine a bitter taste. The palm wine tappers of the Eastern States of Nigeria and the Cameroons claim that the bark retards the tendency of the wine to become sour. Inhibitory effect of the extracts of *Saccoglottis gabonensis* have been extensively studied by Faparusi & Bassir (1972) and Okafor (1975*a*) and shown to contain a phenolic principle which lowers the titratable acidity levels in palm wine. Furthermore, the extract inhibits the growth of *Sarcina lutea* (Okafor, 1975*b*) but had no inhibitory effect on the yeast population of the palm wine. Preservative effect of *Saccoglottis gabonensis* may therefore be in its ability to inhibit the growth of the acid-producing bacteria in the palm wine. With increasing urbanization, the need for an effective method of preservation of palm wine has become urgently necessary.

In an attempt to develop a safe and simple method for the preservation of palm wine, some investigators have determined the chemical composition and the microbial content at various stages of fermentation. Chinnarasa (1968) determined the chemical composition of fresh sweet palm wine. The results are shown in Table 2.

Table 2. Chemical composition of fresh sweet palm wine

Specific gravity at 30°C	1.057
Total solids	16.0%
Sucrose	13.0%
Ash	0.36%
Protein	0.36%
Vitamin C	10 mg/100 ml
Alcohol	nil

Data compiled from Chinnarasa (1968)

Palm sap in situ is essentially a sugar solution having a sugar content ranging from 6.00 to 13.00 g per 100 ml (Ekundayo & Faparusi - unpublished work). Six different sugars including raffinose, sucrose, maltose, melibiose, fructose and glucose have been identified. As fermentation progresses, there is a decrease in the sugar content, various organic acids and alcohols are formed, and a soury taste develops (Table 3).

The preservation and bottling of palm wine Investigations on the preservation of palm wine have been confined to the use of chemical preservatives e.g. sulphites, benzoic acid and sorbic acid (Okafor, 1975*a*). A concentration of sodium

Table 3. Progressive changes in the taste, alcoholic content, sugar and acid content of fermenting palm wine

Hours after collection at 28 - 30°C	3	6	12	24
Taste	Fresh and sweet	Sweet	Slightly Sour	Very sour
Alcohol (%)	3.78	4.84	6.32	6.70
Sucrose (%)	6.80	4.18	1.48	0.35
Acetic acid (%)	0.49	0.54	0.57	0.69

Data compiled from Chinnarasa (1968)

metabisulphite containing 350 parts per million of sulphur dioxide was necessary to arrest fermentation of palm wine. Unfortunately at this concentration, the taste of sulphite was noticeable. Also, the amount necessary to arrest fermentation exceeds the limit (70 p.p.m.) allowed for alcoholic beverages. Benzoates on the other hand are not permitted in alcoholic drinks under the UK Food and Drugs Act and the American Federal Food, Drug and Cosmetic Act of 1938.

Okafor (1975*a*) reported that sorbic acid inhibited to varying extents microorganisms in palm wine but it was, however, not an entirely satisfactory preservative. Investigations have continued on the preservation of palm wine, so as to retain most of its natural flavour.

Ingenious experiments by Chinnarasa (1968) involved the subjection of palm wine in 20 oz bottles to different temperature treatments, between 50 - 90°C, by means of a water bath. A temperature of 50°C for one hour was found to be adequate in arresting fermentation. The method generally used now is to fill a 20 oz. bottle which is then immersed for 30 min in a slowly stirred water bath held at 68°C. Sweet palm wine of more than 1.025 gravity should not be bottled by this procedure, as the gas pressure caused by the vigorously fermenting palm wine may cause the bottle to burst.

As a result of thermal killing of yeasts by this procedure a sediment is formed on standing. This sediment might be resuspended, if desired, by shaking the bottle. Chilling may also prolong the shelf life of the palm wine. There may be a tendency for the sediment to darken in colour after standing for some months probably due to oxidative changes. The incorporation of sodium metabisulphite within the prescribed limit of 70 p.p.m. of SO_2 in the palm wine before pasteurization appears to arrest the oxidative changes.

The economics of bottling palm wine There are no official statistics to indicate the extent of the trade carried out internally with this commodity. Information obtained from local sources suggests that the amount of palm wine consumed is more than three times the amount of beer consumption. On this basis, it is estimated that about 225 million litres of palm wine are consumed annually in Nigeria alone, and probably another 110 million litres used for the preparation of crude spirit.

Palm wine is consumed by people of low and middle income groups. Since these groups constitute over 90% of the population, a flourishing trade could develop through adequate bottling and distribution of the bottled product.

Alcohol production from palm wine

The alcoholic beverage prepared by crude distillation of fermented sap of raphia palm *(Raphia hookeri)* is popularly known in Nigeria as 'Ogogoro'. Other local names exist for it in other parts of Nigeria. In Ghana, it is referred to as 'Apeteshi'.

Ogogoro is naturally colourless but often it is marketed in various colours and has a variable alcohol content as shown in Table 4.

Table 4. Alcoholic content, pH, and residue of some samples of Ogogoro from Nigeria

Source of product	Alcohol content (%, w/v)	pH	Sample residue (%,w/V)
Erin (colourless)	34.0	4.5	0.04
Ikorodu	26.8	4.2	0.11
Ikenne	35.2	4.4	0.00
Akoka	39.9	4.6	0.14
Erin (brown)	35.8	4.1	1.76

Odeyemi, J. Unpublished paper presented at the Fifth International Conference on Global Impacts of Applied Microbiology, 21-26 November 1977, Bangkok, Thailand

The major areas of Ogogoro production in Nigeria are the riverine areas of the Southern States where some 45,000,000 litres are produced annually. Its consumers are mainly the low income earners.

Ogogoro plays a traditional role in that it is considered a special drink at rituals and it is a traditional solvent for many medicinal preparations.

Method of preparation The important stages in the preparation of Ogogoro are: (a) tapping of palm wine; (b) fermentation; (c) distillation; and (d) bottling.

Palm wine, the basic ingredient for Ogogoro production is first tapped using the conventional method described by Faparusi (1973). The sap is pooled in a metal drum and fermentation allowed to take place for some 24 h with occasional stirring to aid fermentation. The fermented sap is put into the still and distilled. Usually the first condensate is discarded. The distillate obtained is often redistilled to produce a liquor with a higher percentage alcohol content. The rectified liquor is stored in 200 ℓ metal drums for bottling.

The following factors affect the quality of the end-product: (a) the temperature of fermentation; (b) types of yeast and other microorganisms involved in the fermentation; (c) the pH of the sap; (d) oxygen requirement; (e) stirring; and (f) duration of fermentation.

Fermentation temperature is usually between 25°C and 30°C, which is the normal tropical ambient temperature. The yeast strains involved are the wild types and the right fermentation pH level is 4.0 to 4.5. The duration of fermentation is highly critical. For good quality Ogogoro, the maximum period of fermentation time confers a very strong and unpleasant smell to the product.

Vinegar production from palm wine

Vinegar has traditionally been produced from a variety of raw materials including fruits (e.g. oranges, grapes, pineapples and bananas) and from cassava, yams, maize and guinea corn. Apart from the traditional raw materials, palm wine has become increasingly important as a raw material for the production of vinegar. The conventional method of production from palm wine involves continuous fermentation of the raw material (palm wine) in partially filled barrels until the product eventually changes into vinegar (Kuboye, 1977).

The microorganisms involved in the fermentation process are the indigenous microflora of the palm wine. These include *Shizosaccharomyces pombe*, *Saccharomyces cerevisiae*,

S. chevalieri, *Leuconostoc mesenteroides* and *Lactobacillus plantarum* (Bassir, 1962; Faparusi, 1966, 1971, 1973; Okafor, 1972, 1975*a*,*b*).

These microorganisms convert anaerobically the sugar contained in palm wine to ethyl alcohol and acetaldehyde which are subsequently oxidized to acetic acid by the vinegar bacteria. These are usually members of the genus *Acetobacter*. The species involved include *A. aceti*, *A. pasteurianum* and *A. xylinum*.

The pilot scale plant, designated the generator, was recently designed at the Federal Institute of Industrial Research, Oshodi, Nigeria for production of vinegar. The generator comprises of three compartments. The first compartment is the distributing chamber into which the fermented palm wine flows in a small stream. The middle compartment contains the filler material which is normally corn-cobs, while the third is the receiving chamber. The generator is initially treated with unpasteurized vinegar in order to acidify and impregnate the filler material with vinegar bacteria. Palm wine of initial acidity of 1.7% is then run through the generator at the rate of about 20 ℓ per hour at a temperature of 27-30°C. The development of acidity is monitored by intermittent analysis of aliquot samples from the generator every 24 h.

Fresh made vinegar is somewhat harsh in flavour and odour. The flavour is attributed to the presence of higher alcohols, acetaldehyde and acids. The harsh flavour eventually disappears and it is replaced by a mild agreeable odour. Vinegar is kept in well filled barrels or tanks and allowed to stand for some months.

For vinegar to be attractive, it should be brilliantly clear. This is accomplished by filtration. However, after filtration, vinegar becomes cloudy due to microbial growth. This is prevented by pasteurization which is carried out by immersion of filled bottles of vinegar in water heated at 60°C for 15 min.

The palm vinegar produced has been tested along with other imported varieties for acceptability. Many consumers rate the palm vinegar as being reasonably mild and it possesses an attractive colour.

Economics of vinegar production Adequate supply of a cheap raw material, palm wine, is available in Nigeria. It is estimated that about 450 million litres of palm wine are produced annually in the Niger delta area of Nigeria. Consequently, vinegar production from palm wine has some market potential and will help to supplement the supply of dilute acetic acid which is required as a rubber coagulant in the Nigerian rubber industry.

The production of pito and burukutu beer in Nigeria and Ghana

The preparation of traditionally fermented alcoholic beverages from various cereals is widespread in Africa (Ekundayo, 1969; Faparusi, 1970; Van der Walt, 1956). These beverages are of immense social and nutritional significance.

Those produced in Nigeria and Ghana on which intensive studies have been carried out include pito and burukutu.

Pito The traditional method for preparing pito, a fermented beverage from maize, sorghum or a mixture of both was described by Ekundayo (1969) and the microbiological aspect of the process was investigated.

A widespread procedure of pito preparation among the Bini tribe in Nigeria is to wash and soak cereal grains (maize, sorghum or a combination of both) in water for 2 days after which they are malted by leaving for 5 days in baskets lined with moistened banana leaves. The malted grains are ground, mixed with water and cooked. The mash is allowed to cool and is filtered through a fine mesh basket. The filtrate is then left usually overnight until it tastes slightly sour and is boiled to concentrate. A small quantity of the "starter" (sediment from a previous brew) is added to the cooled concentrate and left overnight. The product is pito, a dark brown liquid with taste varying from sweet to bitter.

The microorganisms involved in all stages of pito production were investigated and isolated by Ekundayo (1969). Those present on the surface of the cereal grains during steeping and malting were isolated in pure cultures and identified. They include *Rhizopus oryzae*, *Aspergillus flavus*, *Penicillium funiculosum*, *P. citrinum*, *Giberella fujikuroi* and *Botryodiploidia theobromae*.

The microorganisms involved in converting the sweet mesh extract to sour and those in the "starter" responsible for the ultimate conversion of the extract to pito have also been investigated and these include *Geotrichum candidum*, *Candida* spp. and *Lactobacillus* spp. which were frequently isolated and *Aspergillus vesicolor*, *Penicillium simplicissimum* and *P. purpurogenum*, which occurred occasionally.

Generally, fermentation can only occur when some of the starch present in the cereal grains has been converted into sugars. In pito, this saccharification is achieved by means of 2 days soaking in water followed by leaving for 5 days to germinate as previously mentioned. Similar malting procedures have been described in Kaffir beer production (Williamson, 1955; Platt, 1964). It has been maintained that the saccharification process is due to fungal amylases, mainly produced by *Aspergillus flavus* and *Mucor rouxii*

(Platt, 1964). An opposing viewpoint is that fungal activity is unimportant and that the cereal amylases are the effective agent in saccharification (Van der Walt, 1956). As Platt (1964) investigated traditional methods of making Kaffir beer from maize in Malawi and Novellie and co-workers, its industrial production from sorghum in South Africa, their findings did not necessarily conflict with that of Platt (1964). Later studies by Ekundayo (1969) revealed that both fungal activity and cereal amylases may be important in bringing about saccharification in pito production.

The first phase of the pito (and of Kaffir beer) fermentation is souring, lactic acid production by organisms from the atmosphere mainly lactobacilli being responsible.

The lactobacilli bringing about the souring phase in the industrial production of Kaffir beer have been studied by Van der Walt (1956); thermophilic species were particularly important. Pure cultures of lactobacilli are not yet employed even in the industrial process, at most a 'starter' from a previous batch is used (Schwartz, 1956). It is the lactic acid produced in the first phase of fermentation which distinguished Kaffir beer and related beverages from European beers.

The "starter" employed in the second phase of pito fermentation has been shown to contain mainly *Candida* spp. This observation contrasted that of Van der Walt (1956) in which *Saccharomyces cerevisiae* was shown to be the predominant yeast in alcoholic fermentation.

Pito has been shown to resemble Kaffir beer in containing lactic acid, sugars, amino acids and about 3% alcohol. Studies on the vitamin content are yet to be undertaken but pito will presumably resemble Kaffir beer which contains abundant thiamine, riboflavine and nicotinic acid (Aucamp *et al.*, 1961) and perhaps significant amounts of ascorbic acid, although the vitamin content of the industrial product has recently been deteriorating. Like Kaffir beer, pito deteriorates rapidly on storage and within about 48 h of preparation it becomes undrinkable.

Burukutu Beer Burukutu is an alcoholic beverage brewed from guinea corn grains (*Sorghum vulgare, S. bicolor* Moench) in the Savannah regions of Nigeria, Republic of Benin and to a lesser extent in Northern Ghana.

The following procedure is based on the report by Faparusi (1970). Sorghum grains are steeped in water overnight. The grains are then poured into a basket and the water is allowed to drain off for about 2 h.

The grains are spread on leaves in a bed of about 10 cm thickness and covered with another layer of leaves. During this malting period, the grains are watered on alternate days and occasionally turned over.

Germination, which starts within 24 h after steeping, is allowed to continue for a period of 4-5 days. The length of plumule is used in judging when germination has gone far enough.

After germination, the malt is spread in thin layers in the sun to dry for 1 to 2 days. The dried malt is ground into powder. In the next stage 'gari' - a starchy powder produced from the tuber of the cassava plant, *Manihot utilissima* Pohl - is added to a mixture of the ground malt and water and stirred with a stick. The resulting mixture (gari-malt powder-water) is roughly in the ratio 1:2:6 by volume. The mixture is left to ferment for 2 days and boiled for about 4 h. The drink is then left to mature for another period of 2 days. The end-product is a suspension of some particles in a creamy liquid.

Faparusi (1970) determined the different types of sugars in burukutu beer. The brew contained large amounts of different reducing and non-reducing sugars (Table 5).

Table 5. Analysis of sugars in Burukutu beer. The samples were analysed immediately after the two-day maturing period

Sugar	Concentration (%) by weight
Fructose	0.20
Mannose	0.65
Glucose	0.60
Galactose	0.36
Galacturonic acid	0.48
Sucrose	0.20
Maltose	0.42
Isomaltose	0.38
Raffinose	0.60
Maltotriose	0.56
Stachyose	0.43
Maltotetraose	0.16

From Faparusi (1970)

The malting process in burukuto preparation has been studied. Von Holdt & Brand (1960) reported that only fructose, glucose and sucrose were present in ungerminated sorghum grains. They also showed that maltose, isomaltose, maltotriose and traces of higher malto-oligosaccharides were present in germinated grains. Nordin (1959) also reported the presence of stachyose in sorghum grains. Faparusi (1970) found that ungerminated sorghum grains contained varying concentrations of sucrose, glucose, fructose, raffinose, maltose, maltotriose, isomaltose and stachyose. He observed changes in the concentrations of the sugars during malting.

The purpose of malting is to effect the hydrolysis of starch to fermentable sugars. It is the view of many workers that amylases are responsible for this process of hydrolysis. There have been conflicting views as to when the activity of these enzymes becomes manifested. Novellie (1966) concluded that the enzymes are synthesized during the germination stage. However, Dulbe & Nordin (1961) reported that the amylases are usually present in the insoluble condition in the grains even before germination begins.

During the steeping stage, the metabolic activities of the grains which were previously dormant were resumed. Some of the sugars would be used in the metabolic processes. Sorghum grains germinate rapidly and hence will require high energy-level compounds. This could be responsible for the initial fall in the concentrations of some of the sugars reported by Faparusi (1970).

However, when the amylase activity resumes, more sugars will be produced than is required for metabolism, thus the concentrations of these sugars increase.

The microbiology of burukutu beer was studied by Faparusi , Olofinboba & Ekundayo (1973). The microbial flora types of both home and laboratory brewed burukutu beer were found to be the same. Over ten genera of microorganisms were cultured from the sorghum malts used in the preparation of burukutu beer. Apart from the five yeast species - *Saccharomyces cerevisiae, Candida tropicalis, C. krusei, Hansenula anomala* and *Kloeckera apiculata* - five species of moulds were isolated. These included *Mucor rouxii, Aspergillus flavus, A. oryzae, Penicillium citrinum,* and *Oospora* spp. *Lactobacillus* and *Streptococcus* spp. were the two bacterial types found in the malts.

When the mixture of mash and gari (a starchy food substance produced from cassava plant (*Manihot utilissima*) has been allowed to ferment, the microorganisms isolated consisted of such fast fermenting yeasts as *Saccharomyces cerevisiae* and *S. chavalieri*. The commonly encountered bacterial species included *Leuconostoc mesenteroides, Lactobacillis* spp.

(*L. brevis*, *L. fermenti* and *L. delbrukii*) and *Streptococcus lactis*.

During the process of fermentation at room temperature (28-30°C) the pH of the fermenting mixture usually fell from around 6.4 to about 4.2 within a period of 24 h and became 3.7 after 48 h.

The microflora of the burukutu beer undergoing maturation have also been studied; it includes the yeasts *Candida mycoderma*, *C. tropicalis*, *Hansenula anomala*, *Kloeckera apiculata*, and *Saccharomyces pastorianus*. The yeasts usually formed pellicles on the clear supernatant of the beer if left unconsumed for a period of up to 6-7 days. Three species of *Acetobacter (A. capsulatum, A. rancens* and *A. viscosum)* were isolated from the maturing beer. Other bacterial species isolated were *Streptococcus mucilaginosus*, *Lactobacillus pastorianus*, and *Pediococcus cerevisiae*.

Similarly, microorganisms were cultured from samples of gari used in the preparation of burukutu beer. These included *Aspergillus flavus*, *Mucor* spp, *Candida mycoderma*, *C. guilliermondii*, *Oidium lactis*, *Lactobacillus fermenti* and *L. brevis*.

The yeasts and moulds isolated from the malts are not unexpected in view of the humid atmosphere in which the grains were malted. Also, during malting, heat is generated within the heap of the germinating grains. This increase in temperature would aid growth. Most of the microorganisms found on the malts have diastatic ability and will therefore contribute to diastatic conversion of the carbohydrates in the germinated grains.

Mucor rouxii and other moulds are important for their diastatic activity (Van Der Walt, 1956). The occurrence of amylolytic microorganisms on cereal products is widespread. Webg (1945) concluded that the diastatic activities of mould flora on maize and sorghum accounted for the bulk of sugar production required to attain the alcoholic content in Kaffir beer.

The hetero- and homo-lactic acid bacteria found in the fermenting mixture of mash and gari ferment sugars to lactic and acetic acids in addition to the formation of ethanol and evolution of carbon dioxide. Activities of these organisms are also responsible for the fall of pH to 4.2 within 24 h of fermentation.

The *Acetobacter* spp and some yeasts found in the beer during the stage of maturation are mainly spoilage organisms. The *Acetobacter* possesses oxidative dissimilation ability and oxidizes ethanol to acetic acid, thereby increasing the vinegary flavour of the beer. Laboratory and home-brewed burukutu contains about 0.4 to 0.6% acetic acid.

To the consumers, it is the vinegary odour or taste which determines the quality of the beer. Thus a burukutu beer which is not allowed sufficient time to mature does not attract sufficient consumers.

FERMENTED FOODS

Fermented products of cassava (Manihot utilissima Pohl)

Gari One of the staple foods of the rain forest belt of Western Africa is gari, a granular gelatinised starchy food product prepared from the root of the cassava plant (*Manihot utilissima* Pohl). Gari is consumed regularly as a staple food by more than 30 million of the population in Southern Nigeria, and contributes to 60% of the caloric intake (Akinrele *et al.*, 1962).

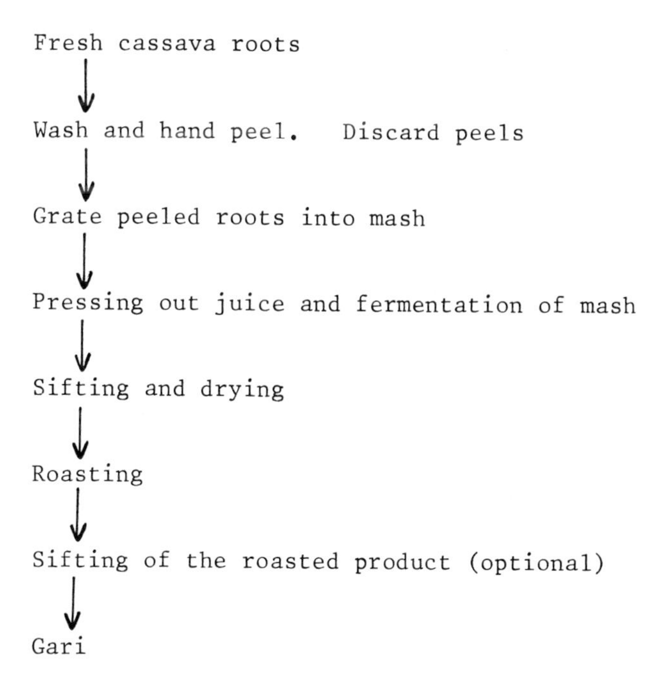

Fig. 1. Flow diagram of village processing (from Onyekwere & Akinrele (1977) Paper presented at the Symposium on Indigenous Fermented Foods, in Bangkok, 21-26 November 1977.

The process flow chart for village gari processing is presented in Fig. 1.

Cassava roots between 1-2 years old before harvesting are

preferred for gari production. The roots are washed and peeled by hand with sharp kitchen knives to remove the outer corky layer and the thick inner cortex. Peeling loss varies between 20-35% and is higher for smaller and irregularly shaped roots. The peelings are discarded while the peeled roots are grated into pulp using the rough side of a perforated aluminium sheet fixed on a rectangular wooden frame. Up to 23 kg per hour are grated. Nowadays, grating in a central place in the village has become popular and this is done by means of a motor-driven pulping machine.

The grated pulp is filled in flour sacks, tied securely and weighed with heavy objects, usually rocks, to facilitate removal of the cassava juice (and incidentally part of the root starch). During this time, fermentation of the cassava mash takes place. Fermentation may be carried on for 2-5 days, depending on how sour the final product is desired.

The fermented pulp is semi-dried to about 60% moisture, By means of local sieves fabricated from palm fronds, fibres and ungrated lumps of cassava are retained and discarded. The pulp is afterwards roasted on a shallow coast iron pot about 50 cm in diameter, over an open charcoal fire. Though the cassava starch is gelatinized at 80°C the pot could reach a temperature of 120°C. The mass is stirred constantly to prevent scorching and to break up the lumps. Roasting is continued until the product becomes dry. The final product when cool is cream coloured and contains about 20% moisture. If yellow gari is preferred, a pinch of palm oil is added during the roasting process.

Mechanized Gari processing The traditional processing of cassava for gari production is evidently beset with the disadvantages of being time consuming, the quality is not unform, the gari has a low shelf-life of about 4-6 weeks and production can no longer keep pace with the food requirements of the urban populations. In realization of these problems, the Federal Institute of Industrial Research, Oshodi, Lagos, Nigeria, pioneered the research in the fermentation of cassava and the subsequent mechanized process using an upgraded village method.

The fully mechanized technique is the product of collaborative effort between the government research centre (FIIR) and a British firm, Newell Dunford Engineering which now manufacture and markets the machine.

This machine is characterized by a relatively sophisticated scientific and technological input. In the FIIR case, the science input was founded upon considerable basic research, whereas with Newell Dunford, a long history of industrial expertise in the drying field was brought to bear on the process. The Mark III Gari plant is now an imported technology

The first of these machines was installed in the Gambia in 1973. The description of the machine together with its functioning was made by Akinrale (1964).

The special areas of interest in the working of the machine are as follows:

1. Cassava roots are peeled by a peeler, which is a rotating drum of concrete mixer design. The combined action of the abrasive lining and the roots rubbing against one another while the drum is rotating accomplishes the peeling effect.

2. The grating machine pulps the roots at a fast speed. At least 70% of the mash when partially dewatered should be retained on sieve of about 6 mm apertures.

3. Fermentation time is reduced from 5 days to 24 h by mixing fresh pulp with a 3-day old fermented cassava liquor at the rate of 1 litre of liquor to 45 kg of mash. The liquor is said to contain a mixture of microorganisms in their early stationary phase.

4. Fermentation is carried out in deep silo with smooth acid resistant inner surface and a conical spout through which the liquid drains, gases escape and from which samples for quality control can be taken.

5. When the pH of the fermenting cassava mash is reduced to 4.0 the desired sour flavour and characteristic aroma would be attained in the gari (Akinrele, 1964).

6. Fermentation seems to proceed best at 55°C in sunlight, and frequent mixing of the fermenting pulp accelerates fermentation. Air, oxygen, and metallic iron are detrimental (Akinrele, 1964).

7. Because of the possible risk of cyanide poisoning, the process plant should be well ventilated. The minimum factory space recommended is 25.5 cubic metres to one kilogram of fresh pulp, unless fume extraction is employed.

8. Fermented pulp is dewatered in a basket centrifuge and the cake is disintegrated in a granulator to remove the thrash. It is roasted in a rotary Kiln (called the garifyer of Newell Dunford, London design). The garifyer is externally heated to 250-280°C by a baffle of hot air. The mash is finally dried to a hot harsh cake of 80%

moisture in a directly fired louvre drier connected to the garifyer.

The gari is milled and packaged in waterproof polythene bags. The industrially prepared gari has a shelf life of at least one year. Many small scale industrial concerns have adopted the above method to produce gari for commercial purposes.

Fermentation of cassava for gari production The microbiology of the traditional preparation of gari has been investigated so as to determine the conditions needed for its mechanized production. The earliest work in this line was done by Collard & Levi (1959), followed by Akinrele (1964) and Okafor (1977).

It is well known that the cassava root contains a cyanogenic glucoside (Linamarin) which renders it poisonous if eaten fresh, and that during the traditional methods, there is a holding stage during which the glucoside decomposes with a liberation of gaseous hydrocyanic acid and after which the material is safe to eat.

Collard & Levi (1959) isolated two organisms from grated cassava during the holding stages of the traditional preparation of gari. These are *Corynebacterium manihot* and *Geotrichum candidum*.

Corynebacterium was found in increasing numbers during the first 48 h of fermentation and was then replaced by *Geotrichum*, which by the third or fourth day was the predominant organism. It was therefore suggested that the process of detoxification of cassava root that occurs in the preparation of gari should be regarded as a two-stage fermentation.

In the first stage the *Corynebacterium* attacks the starch of the root with the production of various organic acids, and the lowering of the pH causes spontaneous hydrolysis of the cyanogenic glucoside with the liberation of gaseous hydrocyanic acid. When sufficient organic acids, including lactic acid, have been produced conditions become favourable for the growth of *Geotrichum* and this then proliferates producing a variety of aldehydes and esters. These last products appear to be responsible for the characteristic taste and aroma of the gari.

Okafor (1977) subsequently isolated a number of other organisms including *Leuconostoc* spp *Lactobacillus* spp and *Alcaligenes*. He also demonstrated the presence of *Corynebacterium* spp and *Candida* spp. He observed that *Leuconostoc* spp were the most frequently occurring organisms. *Alcaligenes* spp though lower in number, occurred briefly on the first two days of fermentation while yeasts were present in increasing numbers.

The abundance of lactic acid bacteria in the fermenting cassava pulp as observed by Okafor (1977) may be explained at least partially by the availability of free fermentable sugar in the tuber. The breakdown of the chemically stable linamarin would appear to be due to the action of the endogenous enzyme, linamarase, of cassava which is released during the grating of the root.

Fermented products of Maize (Zea mays)

Ogi Maize *(Zea mays)* is a major staple food widely eaten in Africa in the form of a sour meal. In the Yorubaland of Nigeria, the first native food given to the weaning babies is called "OGI" - a sour beverage prepared from white maize flour. Similar maize preparations in Ghana are referred to as "Akasa" and "Kenkey". Although Ogi can also be prepared from fermented sorghum *(Sorghum vulgare)* and millet *(Pennisetum typhoideum)* the choice of grain depends on the preference.

Traditional methods of preparation The traditional method of Ogi preparation was described by Akinrele (1966). Essentially, the basic ingredient for Ogi is comparatively fresh or dry white maize. The maize grains are washed and kept in clean water at about 30°C for a period of 1-2 days. In lukewarm water of about 35°C, the steeping time does not exceed 24 h. The maize kernel is ultimately softened by steeping and this facilitates milling.

The softened grain is wet-milled using a typical disk corn mill into a fine slurry which is subsequently sieved through a wire sieve. The pomace is retained on the sieve and later discarded as animal feed while the starch which has been separated with water is allowed to sediment in a deep receptacle (usually a pot). On standing for about 1-2 days it becomes sour as a result of fermentation. At the end of souring the supernatant (called sweet water) is decanted while the sediment, a starch cake of about 55% moisture is called 'Ogi'.

For consumption, the starch cake is turned into a smooth paste with water in a deep bowl followed by the intermittent addition of boiling water to form a hot porridge. The resultant porridge can be sweetened with sugar and consumed as pap. Alternatively, Ogi could be boiled to a thick paste wrapped in leaves and allowed to cool and set to a gel. This is referred to as 'Eko' or 'Agidi'.

Microbiology of Ogi fermentation Akinrele (1966, 1970) carried out intensive studies on the fermentative and biochemical changes occurring during the production of Ogi. This author contended that steeping of maize in water brings about the natural selection of the desirable organisms. He further observed that the fungi *Cephalosporium* spp, *Fusarium*, *Aspergillus* and *Penicillium* were all eliminated during the early steeping period. After 24 h of steeping, the dominant microbes found in order of succession were *Lactobacillus plantarum*, *Aerobacter cloacae* and *Corynebacterium*. The steeping period ends when the steep liquor has been reduced to a pH of about 4.3. This gives rise to a condition favourable for subsequent souring of the corn mash by *L. plantarum* and the yeasts (*Saccharomyces cerevisiae*, *Candida mycoderma* and *Rhodotorula* spp). In practice, therefore, fermentation of maize can be accelerated by priming with previously fermented corn liquor or allowing fermentation to occur as a result of contaminant microorganisms on the grains and from the atmosphere.

The microorganisms cause changes in the maize mash and lactic and acetic acids are produced. These acids bring about change in colour and taste of the mash.

Flavour development is another major result of fermentation in the production of Ogi. The yeasts *Saccharomyces cerevisiae* and *Candida mycoderma* were always able to produce an acceptable flavour even at relatively low acidity due to lactic acid production.

The production of Soy-Ogi Soy-Ogi is a newly developed infant weaning complete protein food in Nigeria which has helped a lot to combat the prevailing high incidence of malnutrition. The production of Soy-Ogi was an exclusive achievement of the Federal Institute of Industrial Research, Oshodi, Lagos, whose main objective was to develop a cheap but satisfactory protein food that might help to alleviate the high incidence of protein malnutrition in Nigerian children, by improving the nutritive quality of the locally prepared Ogi.

The resulting product, Soy-Ogi, is a mixture of 70 parts of corn to 30 parts of soy-beans (used for its protein quality and minerals), is fermented and spray-dried. The food is further enriched with vitamins to increase the nutritional status. The protein value of this enriched product is about 70% which is the recommended allowance for a complete protein food for infants by the Protein Advisory group of the WHO/FAO/UNICEF.

Industrial production The process of manufacture of soy-ogi shown in Fig. 2 consists of preliminary washing of the corn grains in water followed by steeping for 24 h at room temperature (30° - 32°C). The corn is then wet-milled in a disk grinder and sieved with the steep water through a screen mesh containing about 200 μm diameter pores. Soy-beans are cleaned, heated to a temperature of 110°C for 2 h to soften the grains and deactivate the lipoxidase enzyme. The beans are cooled, wet-milled and sieved by means of the screen mesh. The corn and soy slurries are mixed in a dry weight proportion of 70 to 30 and left to ferment as a result of the microflora developed during the steeping phase of the corn. Soya has been found to have a strong accelerating effect on the fermentation reducing the period from 72 h in the case of plain Ogi to 3 h for Soy-Ogi (Akinrele *et al.*, 1970). After fermentation the product is fortified with vitamins and minerals, sugar to taste, spray-dried, sieved through a screen mesh and packaged.

Soy-Ogi is highly acceptable as it resembles Ogi in the traditional infant food both in flavour and in physical characteristics. No colour change or development of flavours occurred on storage for over 6 months. It therefore has a relatively longer shelf-life than the traditional Ogi.

Soy-Ogi is rich in proteins (20.3%), amino acids, minerals (1%), fat (6.3%), carbohydrates (63.7%).

The relative cost of Soy-Ogi is far less than for other infant foods. It has been used in treatment of children suffering from Kwashiokor and it is known to minimise the frequency of diarrhoea normally observed with a milk diet.

USES OF FUNGI IN NON-FERMENTED PROCESSES

Waste fermentation

Waste utilisation is an area of biotechnology in Central Africa in which microorganisms, fungi in particular, have been used in obtaining single cell protein and in fermentation of agricultural wastes to produce industrial alcohol and vinegar.

Two familiar examples will be described to illustrate the progress made in this aspect. They are viz. fermentation of coffee pulp to produce vinegar and microbial single cell protein production.

Fermentation of coffee pulp

The possibility of converting coffee pulp into utilizable products like industrial alcohol, potable spirit or vinegar was investigated by Akinrele & Levi (1959). This was in

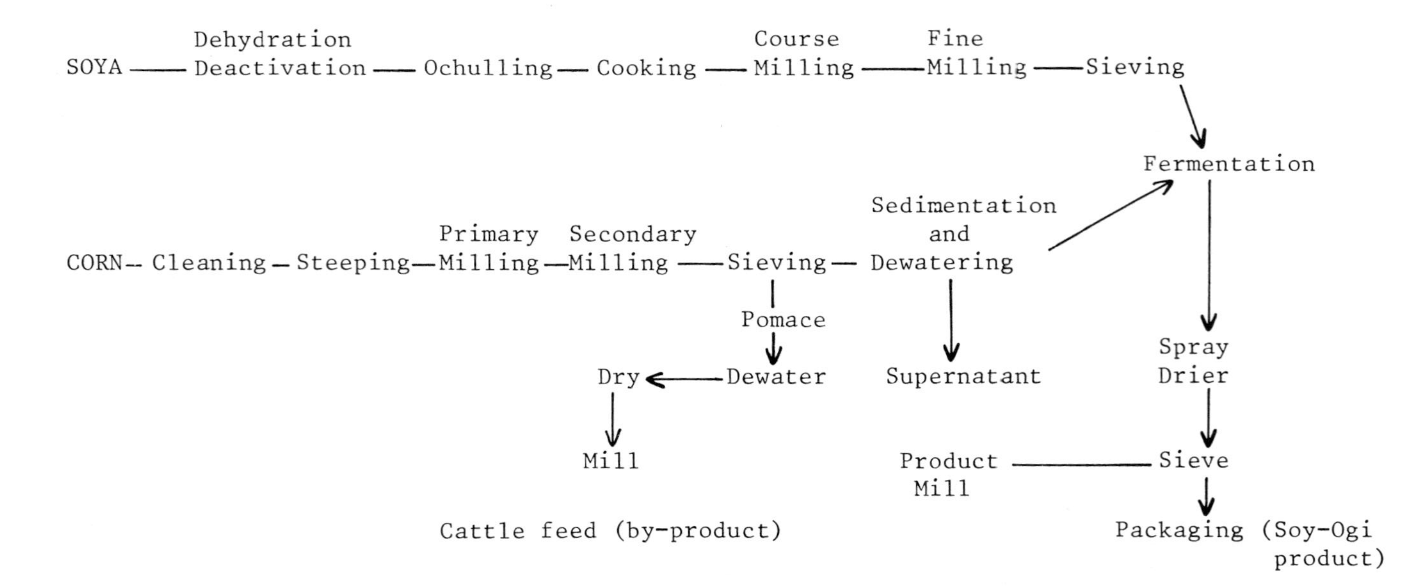

Fig. 2. Flow chart for Soy-Ogi production. From Onyekwere & Akinrele (1977). Paper presented at Symposium on Indigenous Fermented Foods in Bangkok, 21-26 November 1977.

response to a request from the Cameroons Development Agency for advice on the utilization of coffee pulp which is available in considerable quantity at their Santa Coffee estate. Akinrele & Levi (1959) carried out some fermentation trials by inoculating locally available yeasts into pulp samples and incubated for 7 days at 70°C. The yeasts used included *Schizosaccharomyces pombe* (prepared from palm wine), *Saccharomyces cerevisiae* (as brewers inoculum) and *Saccharomyces cerevisiae* (which was reconstituted from dessicated bakers yeast).

From the results of their experiments they concluded that although inoculation with *Schizosaccharomyces octosporus* or *S. pombe* resulted in good yields of alcohol, the low sugar content of coffee pulp (about 7% sugars) would militate against commercial production of alcohol based on coffee pulp as raw material. That notwithstanding, it might be possible to produce potable spirit or wine or vinegar from coffee pulp. Vinegar prepared from such fermentations with *S. octosporus* is reported to possess a clear Rhine-wine colour and a taste resembling that of whisky.

Microbial protein production In the general search for cheap sources of proteins to enrich indigenous staple foods which are mainly starchy, microbial protein production has become important in Central Africa. Protein deficiency diseases exist in about 60% of the populated areas of the world, particularly in tropical countries. It has been realised that it is difficult to supply world dietary protein requirements by conventional agriculture (Strasser *et al.*, 1970). Consequently, foods containing protein from unconventional sources are becoming increasingly important as a source of protein. As a result, much attention has been given to the production of SINGLE-CELL PROTEIN all over the world. In Central Africa, the culture of microorganisms on hydrolysed tropical plant products seems to have particular potential. A crop whose products have been used in pioneering research in Nigeria is cassava.

Protein enrichment by growth of microorganisms in cassava products Very few studies have been carried out in the field of protein enrichment of cassava. However, small scale laboratory experiments have indicated that yeast or fungi could be grown successfully on hydrolysed cassava.

Three different strains of yeasts have been used in the experiments. These include *Candida utilis* which gave a total protein content of 35% , *Rhodotorula gracilis* which gave 26.7% crude protein and *Hansenula seturnus* which gave 15.5% crude protein.

The fermentation technology unit of the Federal Institute of Industrial Research, Oshodi, Nigeria is currently undertaking research of the production of single cell protein from cassava waste. *Geotrichium candidum*, which occurs naturally in fermented cassava mesh (Collard & Levi, 1959), has been shown to offer a potentially edible source of single cell protein. Successes recorded so far include the selection and the determination of the optimum conditions and nutrient requirements of *Geotrichum candidum*, the experimental organism for single cell protein production.

CONCLUSION AND SUGGESTIONS

In the foregoing account, emphasis has been placed on traditional fermentation processes and discussions on brewing industries based on modern biotechnology have been intentionally omitted. The reason for excluding modern wine and beer industries is clear. The biotechnological researches involved in such industrial set-ups are carried out in Europe where the original brand is brewed. In most central African countries, therefore, the local brewing industries use formulations already perfected in the biotechnologically advanced European countries.

The problems enumerated by Hesseltine (1965) as militating against progress in food fermentation technology in the Orient seem to be widespread in developing countries. Since some of the current methods of fermenting foods will continue to be used in developing countries where there are little funds for sophisticated technological research, it is essential that the problems of traditional fermentation processes be identified and solutions found to them.

The problem of taxonomy of the microbial species involved in the various fermentations has become increasingly evident and the validity of some microorganisms involved in the processes is doubtful.

It may be necessary to obtain pure cultures with greater fermentative capabilities (perhaps through hybridization) and that would produce desirable flavours under the conditions of the fermentation.

Because traditional fermentations are by their nature not carried out under controlled conditions, products of variable composition, taste and flavour are obtained. It may be desirable to carry out a survey of the most acceptable form of the product and researches conducted into the best and cheapest method of obtaining such quality product.

Considering the great protein requirement in this part of the world where diet is predominantly carbohydrate, concerted effort should be made in single cell protein production research based on the utilisation of carbohydrate wastes by microorganisms for growth.

REFERENCES

AKINRELE, I.A. (1964). Fermentation of cassava. *Journal of Science, Food and Agriculture* 9, 589-594.

AKINRELE, I.A. (1966). A biochemical study of the traditional method of preparation of "Ogi" and its effect on the nutritive value of corn. *Ph.D. thesis, University of Ibadan, Nigeria.*

AKINRELE, I.A. (1970). Fermentation studies on maize during the preparation of a traditional African starch-cake food. *Journal of Science, Food and Agriculture* 21, 619.

AKINRELE, I.A. 7 LEVI, S.S. (1959). Fermentation of coffee pulp. *Federal Institute of Industrial Research of Nigeria. Research Report No. 4.*

AKINRELE, I.A., COOK, A.S. & HOLGATE, R.A. (1962). The manufacture of gari from cassava in Nigeria. *First International Congress of Food Science & Technology* 4, 633-644.

AKINRELE, I.A., ADEYINKA, O., EDWARDS, C.C.A., OLATUNJI, F.O., DINA, J.A. & KOLCOSHO, O.A. (1970). The development and production of Soy-Ogi - a corn based complete protein food. *Federal Institute of Industrial Research, Nigeria. Research Report No. 42.*

AUCAMP, M.C., GRIEFF, J.T., NOVELLIE, L., PAPENDICK, B., SCHWARTZ, A.M. & STEER, A.G. (1961). Kaffir-corn malting and brewing studies. *Journal of Science, Food and Agriculture* 12, 449-456.

BASSIR, O. (1962). Observations on the fermentation of palm wine. *West African Journal of Biological and Applied Chemistry* 6, 20.

CHINNARASA, E. (1968). The preservation and bottling of palm wine. *Federal Institute of Industrial Research, Nigeria. Research Report No. 38.*

COLLARD, P. & LEVI, S.S. (1959). A two stage fermentation of cassava. *Nature, London* 183, 620-621.

DULBE, S.K. & NORDIN, P. (1961). Isolation and properties of sorghum α-amylase. *Archives of Biochemistry and Biophysics* 94, 121-127.

EKUNDAYO, J.A. (1969). The production of pito - a Nigerian fermented beverage. *Journal of Food Technology* 4, 217-225.

EKUNDAYO, J.A. & FAPARUSI, S.I. (1970). Variation in the microflora, sugar content and tastes of Raphia palm wine during the tapping period. *Unpublished report.*

FAPARUSI, S.I. (1966). A biochemical study of palm wine from different varieties of *Elaeis guineensis. Ph.D. thesis, University of Ibadan, Nigeria.*
FAPARUSI, S.I. (1970). Sugar changes during the preparation of burukutu beer. *Journal of Science, Food and Agriculture* 21, 79-81.
FAPARUSI, S.I. (1971). Microflora of fermenting palm sap. *Journal of Food Science and Technology* 8, 206.
FAPARUSI, S.I. (1973). Origin of initial microflora of palm wine from oil palm trees *(Elaeis guineensis). Journal of Applied Bacteriology* 36, 559-565.
FAPARUSI, S.I. & BASSIR, O. (1972*a*). Factors affecting the quality of palm wine. I. Period of tapping a palm tree. *West African Journal of Biological and Applied Chemistry* 15, 17-23.
FAPARUSI, S.I. 7 BASSIR, O. (1972*b*). Factors affecting the quality of palm wine. II. Period of storage. *West African Journal of Biological and Applied Chemistry* 15, 24-28.
FAPARUSI, S.I. & BASSIR, O. (1972*c*). Effect of extracts of the bark of *Saccoglottis gabonensis* on the microflora of palm wine. *Applied Microbiology* 24, 853-856.
FAPARUSI, S.I., OLOFINBOBA, M.O. & EKUNDAYO, J.A. (1973). The microbiology of burukutu beer. *Zeitschrift für Allgemaine Mikrobiologie* 13, 563-568.
HESSELTINE, C.W. (1965). A millenium of fungi, food and fermentation. *Mycologia* 62, 149-196.
KUBOYE, A.O. (1977). Vinegar production from palm wine. Paper presented at the Symposium on Indigenous Foods, Bangkok, November 21-26, 1977.
NORDIN, P. (1959). *Chemical Abstracts* 55, 8548.
NOVELLIE, L. (1966). Kaffir-corn malting and brewing studies. XIV. Mashing and Kaffir-corn malts; factors affecting sugar production. *Journal of Science, Food and Agriculture* 17, 354-361.
ODEYEMI, O. (1977). Ogogoro industry in Nigeria. Paper presented at the Symposium on Indigenous Foods, Bangkok, November 21-26, 1977.
OKAFOR, N. (1972*a*). Palm wine yeasts from parts of Nigeria. *Journal of Science, Food and Agriculture* 23, 1399-1402.
OKAFOR, N. (1972*b*). The source of microorganisms in palm wine. *Symposium Nigerian Society of Microbiology* 1, 102-105.
OKAFOR, N. (1975*a*). Preliminary microbiological studies on the preservation of palm wine. *Journal of Applied Bacteriology* 38, 1-7.
OKAFOR, K. (1975*b*). Microbiology of Nigerian palm wine with particular reference to bacteria. *Journal of Applied Bacteriology* 38, 81-88.

OKAFOR, N. (1977). Microorganisms associated with cassava fermentation for gari-production. *Journal of Applied Bacteriology* 42, 279-283.

ONYEKWERE, O.O. & AKINRELE, I.A. (1977). Gari - a fermented food from cassava. Paper presented at the Symposium on Indigenous Fermented Foods, Bangkok, November 21-26, 1977.

PLATT, B.S. (1964). Biological ennoblement - improvement of the nutritive value of foods and dietary regimens by biological agencies. *Food Technology* 18, 662-670.

SCHWARTZ, H.M. (1956). Kaffir-corn malting and brewing studies. I. The Kaffir-beer brewing industry in South Africa. *Journal of Science, Food and Agriculture* 7, 101-105.

STRASSER, J., ABBOTT, J.A. & BATTEY, R.F. (1970). Process enriches cassava with protein. *Food Engineering,* Chilton Company, U.S.A.

VAN der WALT, J.P. (1956). Kaffir-corn malting and brewing studies. II. Studies on the microbiology of Kaffir-beer. *Journal of Science, Food and Agriculture* 7, 105-113.

VAN HOLDT, M.W. & BRAND, J.C. (1960). Kaffir-corn malting and brewing studies. VI. Starch content of Kaffir-beer brewing materials. *Journal of Science, Food and Agriculture* 11, 463-465.

WEBB, R.A. (1945). The chemical microbiology of Kaffir-beer. *Biochemical Journal* 39, XLIX.

WILLIAMSON, J. (1955). *Useful plants of Nyasaland.* The Government Printer, Lomba, Nyasaland.

BIOTECHNOLOGY IN MALAYSIA

R.N. Greenshields

Department of Biological Sciences,
The University of Aston in Birmingham,
Gosta Green, Birmingham B4 7ET

INTRODUCTION

Biotechnology, in the form of the traditional fermentation activities, has been well-established in Malaysia for centuries. This is common to most Eastern or Asian countries (Hesseltine & Wang, 1979). However, biotechnology as a modern multi-disciplinary science and technology has only just become significant in this area. Malaysia is a third-world country but has become over the past four years an important and rapidly developing agro-industrial centre. It has an enormous potential and expertise to produce the substrate for biotechnology together with a clear requirement for the application of this science. Therefore, this situation and the significant economic position that Malaysia holds in South East Asia suggests that there is an interesting possibility for biotechnology that could well influence the science itself.

For an expatriate to determine the state of a science in a country is an almost impossible task; but since there is little biotechnology in Malaysia to any significant extent, this presentation can at least highlight the development possibilities where biotechnology is having, and will have, application in the future.

GEOGRAPHY

Malaysia (Tanah Malayu - Malay Land) consists of two areas, East Malaysia comprising Sabah and Sarawak on the North Western tip of Indonesia, Borneo (Kalimantan) and West Malaysia or Peninsular Malaysia which is the southernmost part of Continental Asia. East Malaysia is largely undeveloped and this presentation is primarily to West Malaysia. West Malaysia consists of a peninsula projecting into the China Sea at 100-104°E from the northern narrow point where it joins Thailand at 10°N (at the Isthmus of Kra) where it is only 40-50 miles wide, extending down to the tip some 750 miles south at ½°N of the Equator, where Singapore Island can be found. On the east is the South China Sea and on the West the Malacca Straits. In the middle it is about 200 miles wide and is divided into two parts by a range of mountains some 300 miles long. The Western area is well developed with agriculture, and secondary industry and it has the majority of the population while the Eastern area remains largely primary jungle although it is being rapidly developed.

CLIMATE

Being just north of the Equator and with no point more than about one hundred miles from the sea, Malaysia is typically tropical with high humidity and uniform high temperature. It is under the influence of the South West and North East monsoons but their effects are minimal. There are no true seasons although the North East monsoon during October to February brings more rain to the Eastern areas. Rainfall is about 100 inches per year with average temperatures of 80 to 90°F (28 to 32°C) with sometimes temperatures as low as 70°F (21°C) at night and in the central areas temperatures as high as 100°F (38°C) in the day. In the mountains, which rise up to 7,000 ft., the temperature can fall as low as 60°C (16°C).

VEGETATION

This is mostly rain forest with some seventy per cent primary jungle covered with typical large trees (Dipterocarpaceae) and Epiphytes. Generally in the plains there is good fertile lateritic (volcanic) soil which has a tendency to erosion when exposed, although the coastal region in the West includes clays and on the East coast, bris. Much of the plain areas and low hills are now plantations of rubber, palm oil and cocoa. Where tin mines have been found, large areas are barren due to the tin tailings.

POPULATION

There are about fourteen million people in Malaysia consisting roughly of 55% Malay (Indonesian origin) of Austro-Asiatic race, 30% Chinese, 10% Indian (Tamil), 5% Eurasian and others. Malaysia has one of the highest population increases in the world, some 3% per annum.

POLITICAL

The development of Malaysia reflects successive Imperial dominance by the Chinese, Portugese, Dutch and finally the British with independence from Great Britain some 21 years ago, August 31 1958. Malaysia is a developing third world country but over the last decade there has been a massive increase in commodity agriculture and mineral wealth which has brought their Gross National Product to being close to almost that of a second world country. Industrialisation is proceeding with the aid of much available labour and tax concessions to attract investment, though this is basically secondary industry. It is a leading partner in ASEAN, Assocation of South East Asian Nations, and provides a key role in this area. It is a democracy with constitutional Royalty derived from its separate states and has enjoyed considerable stability since its independence.

AGRICULTURE

A brief consideration of the agricultural situation in Malaysia is necessary to set the scene for biotechnology because here in Malaysia are the potential substrates for fermentation and the necessary conditions for commodity agriculture.

After the initial development of the tin resources, the rubber plant (*Hevea brasiliensis*) was introduced by the British from Brazil and this dominated the accessible areas where the primary jungle could be cleared and roads introduced without difficulty, basically in the South and West of West Malaysia. Pineapple was also successfully introduced into Southern Malaysia but rapidly reached a peak because of competition from the Pacific areas. This agricultural pattern dominated the country for almost fifty years but with the decline of the natural rubber industry some twenty years ago, agriculture took a new direction with world commodity agricultural crops. In particular, with the introduction of the Oil Palm, mature rubber plantations were decreased in size and replanted with oil palm. This was a spectacular success giving an overall increase in quantity of some 10% per year. Palm oil finds a ready market world wide. Cocoa (*liberica*) strain and coffee have also been recently successfully planted and plantation sizes rapidly increased in certain suitable areas which suit these crops.

From the Bank Negara Malaysia report of 1977 we can find the following information on the agricultural crops:

Table 1. Agricultural report of the Bank Negara Malaysia 1977

Agricultural crop	Development	Quantity per annum
Palm oil	Rapidly increasing	1.7×10^6 tonnes Crude Oil*
Rubber	Stable	1.6×10^6 tonnes Dry Weight
Cocoa	Rapidly increasing	30,000 tonnes Beans
Pineapple	Stable	250,000 tonnes Fresh Weight
Coffee	Increasing	10% of World usage
Tea	Stable	Small
Cassava	Local	Small
Rice	Decreasing slightly	1.8×10^6 tonnes
Copra	Increasing	120,000 tonnes
Pepper	Decreasing	30,000 tonnes
Sugar cane	Slightly increasing	50,000 tonnes

* 20% yield from fresh fruit bunches (8.5×10^6 tonnes per annum)

40% of yield gives a kernal yielding kernal oil and seedcake

Malaysia is often referred to as the fertile crescent and could be the granary of South East Asia since application of only small amounts of organic materials and fertiliser produce an abundance of tropical crops and fruits. A new scene that has emerged over the past five years is the wholesale stripping of large areas of primary jungle particularly in the Eastern regions, initially to harvest the big trees and then to put down palm oil, cocoa and rubber plantations. This process is now also happening in East Malaysia. Environmental concern has now reduced this to a more controlled situation and vast National Parks are being considered to preserve some of the primary jungle areas and to conserve renewable timber resources. It is suspected that climatic changes may have already occurred due to these changes.

THE BIOTECHNOLOGY SCENE

The limited agriculture in the early days coupled with low-level development has meant an almost non-existent biotechnology. Moreover, the crops that were available were high in carbohydrate content and low in protein, this coupled with the lack of animal protein has led to a wide variety of fermentation techniques based on the koji principle to be developed producing upgraded foods like tempeh and ragi and flavour extracts such as soy sauce and other soy flavourings. This topic has been adequately reviewed by Stanton & Walbridge (1969); Wood & Min (1975); Hesseltine & Wang (1967) and Pethiyagoda (1978). Fermented drinks have a restricted development in Malaysia since it is basically an Islamic country, although the Tamil Indians are allowed to make Toddy (Pathiyagoda, 1978) (a natural fermentation of coconut inflorescence sap by yeast and bacteria), Sam-su, the distilled spirit of Toddy, is illegal.

However, the recent development of a wider variety of agro-industrial crops, palm oil, cocoa, coffee and pineapple, to replace the rubber has provided a stimulus to biotechnology in a curious way. This development and its associated secondary industry has given rise to water-pollution on a large-scale despite the high rainfall and fast-flowing rivers, threatening the livelihood of the local Malya-Kampang people by polluting the water essential to their communities. The Malaysian Government has followed the current environmental awareness and enacted regulations to combat this problem (Laws of Malaysia, 1977). This has stimulated a wide variety of biotechnological solutions to meet the necessary restrictions imposed.

In addition, the demand for more protein food particularly animal protein with the improved standard of living by an increased population has led to large Government programmes and incentives to produce more animals (cattle, pigs, poultry and goats) together with a development of milk and milk products production.

This in turn has led to the further development of the animal feed industry since Malaysia does not have the necessary immediate crops for these animals. Again, biotechnological expertise to provide suitable materials for this feed are being developed. The pressure for biotechnology and its expertise is increasing and will become more pronounced as the Eastern area of Peninsula Malaysia opens up and as East Malaysia also becomes developed.

Pollution control and Biotechnology

Palm oil processing and refining, rubber processing, cocoa fermentation and pineapple processing are the main pollution offenders and which have become subject to control and restriction.

Palm Oil Processing

'Palm Oil processing is considered to be the worst pollution culprit but the best economic success upon which the development of Malaysia depends.' It was reported by the New Straits Times in November 1977 that 42 Malaysian rivers were unable to support any form of aquatic life and another 16 were polluted to only a less serious extent. In 1977 a yield of 8 million tonnes of fresh fruit bunches (FFB) grown on 2 million acres produced some 1.7 million tonnes of crude oil from 100 mills (ranging in size from 10 to 80 tonnes FFB per hour) worth some 2,000 million Malaysian dollars (approximately £50 million) but producing 6 million tonnes of effluent at 20-30,000 ppm B.O.D.

Since this process is one of the most important agro-industries in Malaysia and its effects are so dramatic, the process is worth some detailed consideration (Fig. 1).

The palm fruit bunch (PFB) consists of a collection of drupes each containing a stone surrounded by a bright orange (carotene-containing) pericarp which has some 20% oil in the parenchymatous cells. This is heat-treated in the bunch form under steam pressure (40 p.s.i.) to destroy the lipolytic enzymes and to soften the fruit. The result is a condensate liquor effluent. The drupes are stripped off the bunch and the bare bunch burned to give an ash which is returned to the plantation as a fertiliser. The drupes are crushed at high temperature to release and separate the oil and the nut removed. The nut is further crushed to give other high quality oils and seed protein cake. A heavy sludge (brown to brown-grey in colour) is produced at about 80°C which is combined with the condensate strippings and sometimes with hydrocyclone washings of this equipment. The combined effluents are sent to a holding tank where further oil may be reclaimed by skimming. About 0.4 to 0.6 tonnes (sometimes 0.8 tonnes) of effluent ensues per tonne of PFB depending on the efficiency of the plant operation. The effluent can be dried and has been used as animal feed since it contains up to 15% protein (Dalzell, 1977). Its consistency is a problem in disposal or use since it tends to set into a semi-gel on cooling and rapidly becomes unpleasantly infected with anaerobic bacteria.

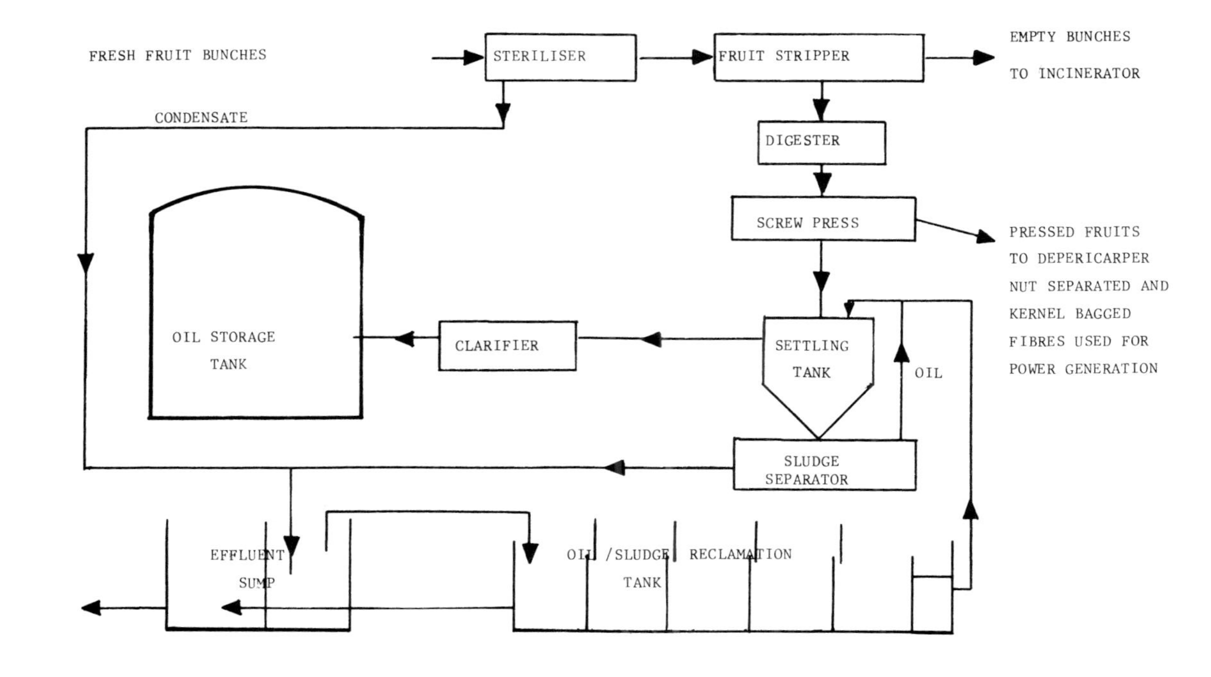

Fig. 1. The palm oil extraction process.

Table 2. Effluent composition of palm oil processing (Etzel, 1977)

pH	4.8 - 5.1 mg 1^{-1}
Oil	6,000 - 11,000 mg 1^{-1}
Total Solids	36,000 - 35,000 mg 1^{-1}
Suspended Solids	10,000 - 28,000 mg 1^{-1}
Nitrogen	500 - 900 mg 1^{-1}
B.O.D.	18,000 - 25,000 mg 1^{-1}
O.O.D.	45,000 - 55,000 mg 1^{-1}
Temperature	80 - 90°C
Flow	0.4 - 0.6 tons ton^{-1} fresh fruit bunches processed

A typical effluent composition is given in Table 2 but basically consists of suspended fruit cells with and without oil, free oil and emulsified oil, sand and dissolved materials both in high and low molecular weight forms. The Government has laid down standards of effluent disposal which came into force in July 1979. These are progressive for watercourse discharge but are basically the highest level standard for land disposal (Table 3). An early conference (Stanton, 1972) led the scene for early research and pollution control research but more recently, three symposia held in Kuala Lumpur by the Oil Plam Growers' Council Technical Committee have considered a wide variety of methods to provide successful solutions to this problem. Figure 2 gives a summary of some of the treatment options considered, some of which have been attempted but largely on pilot-scale only.

Option 1. Solids removal: filtration, decantation, or centrifugation with or without flocculation is a possibility but the quantities involves are large and therefore this is an expensive technique. The filtrate or centrifugate being treated by conventional treatment techniques.

Option 2. Heat assisted total or partial evaporation to give a solid for land disposal, burning or mixing with other materials to provide animal feed. Again, quantities are large and this would be energy expensive.

Table 3. Four generation effluent standards for palm oil mills (Malaysian Division of Environment 1977).

Parameter	standard a 1.7.78	standard b 1.7.79	standard c 1.7.80	standard d 1.7.81
Biochemical Oxygen Demand 3 day, 30°C mg l^{-1}	5,000	2,000	1,000	500
Chemical oxygen demand mg l^{-1}	10,000	4,000	2,000	1,000
Total solids mg l^{-1}	4,000	2,500	2,000	1,500
Suspended solids mg l^{-1}	1,200	800	600	400
Oil & grease mg l^{-1}	150	100	75	50
Ammoniacal nitrogen mg l^{-1}	25	15	15	10
Total nitrogen mg l^{-1}	200	100	75	50
pH	5.0-9.0	5.0-9.0	5.0-9.0	5.0-9.0
Temperature ^{o}C	45	45	45	45

Option 3. Fermentation. The effluent could provide a suitable substrate for yeast or filamentous fungal fermentation to upgrade the solids and fermentable dissolved solids to animal feed (Greenshields, 1977*a*, *b*, 1978). The effluent from the process which is lowered in BOD strength could then be disposed of direct to the land or treated by small conventional effluent treatment systems for either disposal to land or watercourse. Pilot and production-scale units have been tested but the economics have yet to be obtained. However, Malaysia imports some 90,000 tons soya bean meal per year at a cost of 60,000,000 M$ (£15,000,000) and substitution of this would assist balance of payments, and self-sufficiency.

Option 4. Direct Disposal to Land. Experiments on a fairly large scale show this to be feasible though the long term effects are not predictable (Wood, 1978).

Option 5. Standard anaerobic/aerobic sewage treatment with or without methane production in ponds and tanks (Etzel, 1978). This is feasible but solids removal would assist in lowering

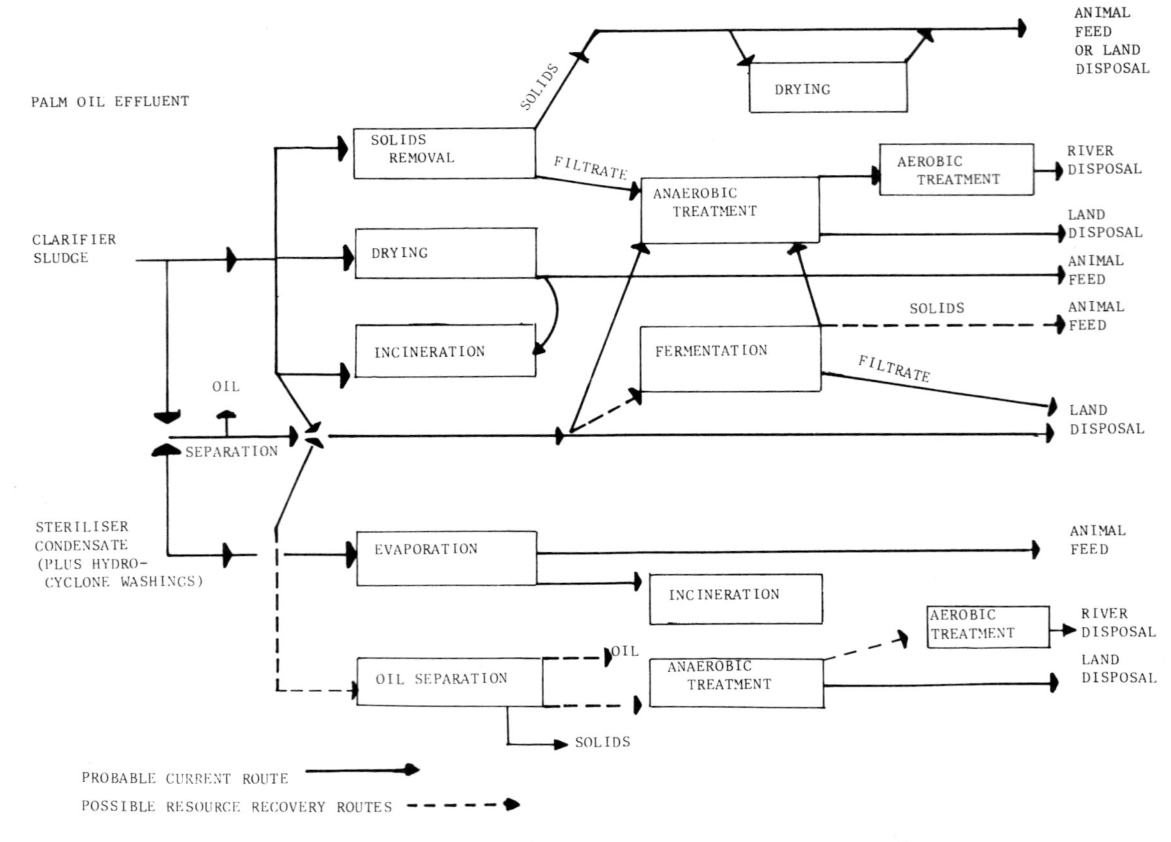

Fig. 2. Palm oil mill effluent treatment options.

the long detention times and scale of land requirements needed (some 20-80 days depending on the system). Large scale trials have been conducted on this with considerable success and make it the cheapest system most likely to be adopted.

Option 6. Oil removal. Solvent extraction of the remaining oil gives a valuable second-grade oil and assists in solids removal (Kirkaldy, 1978). These solids could be useful low-grade animal feed and the removing effluent is easily treatable by standard effluent techniques.

Option 7. A variety of sophisticated techniques such as wet oxidation, reverse osmosis, have also been suggested but quantities are too large for these methods.

All the options are expensive either in energy, capital and operating costs or land area involved. by and large the cheaper methods are not cheap in the long term because they have no return. However, the fines for pollution are also expensive, for example, up to 300,000 M$ per year (£75,000) for a 70 ton FFB/h Mill. Since a licence is necessary to operate a palm oil mill and the granting of this depends on having adequate pollution control to the Government standards, a solution is imperative and one or other of the methods must be adopted.

In the last symposium, the Government (Ho, 1978) suggested that there should now be no difficulty in obtaining the first standard of pollution control and indicated 4 types of system based on biological means, chemical and physical methods, disposal on to the land with or without treatment and direct utilisation; all of which were being tested. 68 out of 128 mills had licences and of these 64% had biological treatment systems, 7% had biological treatment with disposal to land, 7% disposed of effluent with no treatment on the land, 3% had other types of method and 19% had considered no treatment whatever. They felt that biological treatment was the obvious way, basically using anaerobic digestion with aerobic post-treatment a possibility. This was confirmed by the industry although anaerobic ponding proved to be the most popular and cheapest solution, since it operates well in the high ambient temperatures. The use of solids separation as a pre-treatment technique was also a possibility to reduce detention times.

This particular 'hot-potato' of biotechnology in Malaysia is certainly not a completed story by any means and the simple disposal of this technically-valuable by-product is not the only answer. As the population increases and food, energy and land become scarce commodities resource-recovery by biotechnology must become important and play a more serious role in this situation. The accountancy exercise currently conducted to justify present process methods will be re-written with new economic base-lines and rules.

Rubber processing. One of the earliest pollution problems came from rubber processing and derived from the coagulation, separation and washing procedures. These give rise to an ammonium sulphate solution together with proteinaceous materials from the latex juice - an ideal substrate in the Malaysian ambient temperatures for bacterial growth. This, together with a skim serum, form the main effluents. Quantities are not large but the anaerobic fermentation which results from their disposal without treatment gave rise to a high pollution load and an appalling stench in local rivers and streams, widely dispersed throughout Malaysia. These effluents have now been shown to be easily treated by properly controlled anaerobic ponds and the situation is such that the proposed regulations for rubber factory effluents have been advanced in their implementation by a year. However, once again it may not be the entire story since the effluents could be used to supplement the nitrogenous salt requirements of S.C.P. fermentations if such systems were adopted for palm oil effluent treatment.

Pineapple processing. This is another story of neglect now being curtailed by current regulations. Only the inner 40% of the pineapple was used in canning, the remainder, which consists of the outer segments, skin, excess juice and cutting debris, was disposed of into local rivers and thus hopefully out to sea. However, the result was a source of pollution particularly in the tidal reaches of the rivers. These were made so anaerobic that no aquatic life could survive.

An anitial approach to this problem which has been successful is to crush the residues and silage the material for animal feed. The juice which remained was still a problem since it had a low concentration of sugar and was unsuitable for most treatment techniques. Currently, Alagaratnam (1978) has devised a system whereby the juice is concentrated from about 2% up to 6-8% sugar and then fermented to alcohol. This can then be used to make pineapple wine, liqueur or vinegar. The high cost of this process is offset by the high value of the products and the high cost of alternative disposal systems (especially in the particular case in point since transportation was involved). A further possibility that is also currently being considered is the preparation of citric acid, since pineapple not only contains this acid but the juice has considerable potential for this fermentation.

Cocoa fermentation. This crop has only developed in the last five years and has become increasingly important, being one of the major agricultural product growth areas. During the preparation of the cocoa seeds from the pods, the mucilagenous polysaccharides and sugars which surround the seeds ferment rapidly to alcohol and then to acetic acid when exposed to the air. This process is an essential step in cocoa flavour development. In Malaysia the cocoa beans were too acid by comparison with those obtained from Africa and South America. This is probably due to the variety used and an over-ebullience of acetic acid fermentation. Regulation by temperature control and by pressing to remove some of the polysaccharide has succeeded in ameliorating this phenomenon. Nevertheless, the fermentation produces some 25 litres of these 'cocoa sweatings' per tonne of beans in the process and constitutes an effluent. Perhaps even an effluent problem when the cocoa plantations are fully developed. In one company it already amounts to 100,000 gallons per year. Whilst alcohol can be made as a by-product, since the acetification is desired in the cocoa process, the possibility of making a vinegar has been considered (Shepherd, 1977). Such a product has been made and shown to be similar to (5-6% w/v) malt vinegar both in appearance and flavour (Nevantzis, 1978). Although this is acceptable to Europeans, only clear vinegar is acceptable to the Malaysians. Strip-stilling at low concentrations should give a suitable answer.

Future Developments

The fermentation technology developed in the treatment of the effluents has stimulated wider consideration and the following ideas are actively being researched and developed both in Malaysia and in countries which have association with Malaysia. These ideas can only be considered briefly.

Soy Sauce. A considerable simplification of this complex process has been achieved by identifying and isolating the specific organisms concerned and determining the precise fermentation conditions. Pure culture techniques and optimal fermentation then allows the overall process to be reduced in time from three years to three months with consistent product quality. Such improvement in process techniques coupled with the availability of the local substrate could enable Malaysia to develop its own soy-extract industry. However, competition from other nations with their market control at the moment does not easily allow development.

Enzymes. Direct production of enzymes from local crops, for example bromelain from pineapple and papain from payaya provides a clear possibility for the start of an enzyme industry. However, many other available substrates could be used to manufacture valuable enzymes by fermentation that would have an immediate application in the Malaysian agro-industrial sitiatuon. The 'sweet-tooth' of the Malaysians ensures a ready market for sugar extracts of all descriptions and thus the process of enzymic hydrolysis of starch to glucose and its conversion to isomerose (glucose and fructose) by the glucose isomerase enzyme would be an economic industry. The enzymes could be manufactured in Malaysia rather than imported from Europe.

The final extraction of oil from the palm oil sludge leaves much to be desired, up to 2% of the oil can be left bound up in the parenchymatous cells. The use of hemicellulase, cellulase and gum enzymes have been shown to be beneficial in releasing this oil in the hot oil recovery stage before the effluent is discharged. Again these enzymes could be made on-site at the Mill processing plant by quite simple equipment rather than being expensively imported as is proposed.

The position of Malaysia in the ASEAN complex could ensure a sound market and quick development of this allied agro-industrial business using relatively simple technology and expertise.

Yeast. Currently the Mauri Company of Australia is making bread yeast from molasses by an advanced continuous fermentation system. The economics are somewhat marginal due to the substrate, its availability and transport although the cost of cooling and product drying is also high. The application of enzyme technology to utilise cheap starch substrates may improve the economic balance of this process although the situation does emphasise the problem of distance and transport costs in developing countries.

Wine and vinegar. Although Malaysia is an Islamic country, beer is made to suit those who inbibe alcoholic beverages. Wine, however, is not made despite the fact that there is an abundance of fruits that could be used (even pineapple wastes). This is to a certain extent due to the fact that suitable grapes do not grow well in the Malaysian climate and wine is always traditionally associated with the grape. Some work has been done on wine production from pineapple and this could be extended. As previously mentioned, vinegar is mostly imported and since vinegar production depends on the local alcoholic beverage industry, this process has also not been developed despite a clear requirement for vinegar, both as vinegar and in the preparation of sauces. A possibility for a rural-based industry exists here.

Algae culture. Specialised algal culture (specific *Chlorella* spp.) have been grown in the South of peninsula Malaysia utilising CO_2 as substrate for the Japanese human food market where it is a great delicacy. Although it is an expensive process, the product commands an equally high price. Unfortunately, competition from other countries has made the process uneconomic.

Fermented food and feeds. Traditionally Far East nations have a long history of these valuable protein-upgraded products, but apart from extensive work by the Japanese on the rice-koji fermentation and Paddy straw mushrooms, little has been done to develop what could be a valuable biotechnology that has an immediate impact on the rural community. There is the necessity to study the actual microorganisms involved, their ecology and fermentation conditions to enable the processes to be efficiently carried out with the availability of pure cultures and information on the proces conditions. Tempeh (*Rhizopus* spp. on rice and soya), Tempeh bongkrek (*Rhizopus* spp. on tapioca), Ragi, Dhose (yeast on peas) are all possibilities that are available since they are well-recognised and eaten in Malaysia.

Stanton (1978) has developed a 'fish sauce' fermentation to allow fish-wastes collected whilst fishing to be fermented *in situ* in plastic bags. The fermentation is seeded with a culture taken from previous trips. The fermentation is complete at the high ambient temperatures and by this time the fishing expedition is complete, the product can be utilised immediately on return. Basically this is a silage-type fermentation based on the acid-bacteria such as the *Lactobacilli* and provides safe nutritious high protein animal feed free from pathogens. The concept of village-level biotechnology has much to commend it in developing nations but unfortunately has only been taken up by a few. The further concept of SCP production *in situ* particularly utilising waste substrates in cheap, simple fermenters (plastic bags) has a considerable future (Faust & Prave, 1979). Plastic bags have already many remarkable uses in the Far East and this idea could well be developed easily and quickly.

Alcohol. Most developing nations require petroleum for liquid transport fuel and only a few have direct access to these within their boundaries. Malaysia has sea oil supplies but these in common with the rest of the world are a finite resource. (It is estimated that the Malaysian resources on land at present will last about 12 years). Alcohol has clearly been shown to be a valuable addition to this resource particularly where a country is agriculturally-based and can produce the necessary substrates. Brazil is the prime example of a country that has adopted this concept and has developed a massive programme that is well-advanced, although others such as the United States, Australia and New Zealand are following suit (Hall, 1979). Malaysia has access and could develop substrates for this purpose - cassava, tapioca, sago, with adequate land and expertise to implement such a programme. As the technology for the utilisation of cellulose to alcohol becomes available, this could provide yet another incentive since Malaysia has adequate supplies of wood, woodwaste and sawdust.

SUMMARY

Malaysia provides an ideal area for the development of biotechnology because:

It is a centre for the ASEAN countries and the export potential is well established.

It has clear requirements in its development plans for an agricultural base with related agro-industries. The application of biotechnology would give a natural extension providing further employment, education and rural development of these plans. It would assist in overcoming waste, provide a high degree of self-sufficiency with good international trade and bring appropriate technology to its people.

It has the potential and expertise to provide the substrates.

It has a requirement for protein food and animal feed.

However, there are a number of problems in applying such technology that are inherent to developing countries:

The problem of the investment and return from such technology compared to agriculture and land investment. Quicker and higher returns come from land than the related agro-industry despite Government incentives.

The natural political development criteria often overrule scientific considerations, for example use of language.

The problem of developing bureaucracy and its inability to cope with a developing situation especially when overwhelmed by the vast problems encountered or slowed down by the pace of the rural environment.

The differing philosophies. This cannot be a criticism in any way but it is well understood that Asian philosophy and education is different to Western philosophy in many fundamental matters.

The inappropriateness of biotechnology in a developing nations' economy, this is often a matter of priorities. Moreover, for biotechnology to be applied requires import of foreign materials and expertise which may be inappropriate to the country's philosophy or political thinking.

The problem of the vast scale of application can often defeat the initial approach of a new process particularly when the process has been determined within the conditions of a sophisticated environment and is now faced with the conditions and environment totally different and explored. The availability of facilities can also be important here, for example expertise with instrumentation, availability of water, adequate technical control, availability of spares and maintenance, availability of pure water, effect of temperature and humidity.

Finally, the problem of competition from countries, often first world countries, who control the market and the commodities. A perfectly viable process valuable to the country and inherently commercial can be doomed from the outset by open market control by monopoly situations.

This review could never be complete since it is an ongoing story and it can never be thoroughly stated by any one person. Nevertheless, what is certain is that biotechnology will play an important role in the future of our civilisation as does petrochemistry and electronics today - and Malaysia because of its special position can benefit from its application.

REFERENCES

ALAGARATNAM, R. (1978). Personal Communication.

DALZELL, R. (1977). Palm oil effluent as animal feed. *Oil Palm Growers Council 1st Technical Symposium*, Kuala Lumpur, Malaysia.

ETZEL. J.E. (1978). Anaerobic digestion processes for palm oil effluent treatment. *Oil Palm Growers Council 3rd Technical Symposium*, Kuala Lumpur, Malaysia.

FAUST, U. & PRÄVE, P. (1979). Fermenter design for single cell protein production. *Process Biochemistry* 14 (ii) 28.

GREENSHIELDS, R.N. (1977a). Palm oil effluent problems. *Oil Palm Growers Council 3rd Technical Symposium*, Kuala Lumpur, Malaysia.

GREENSHIELDS, R.N. (1977*b*). Palm oil effluent treatment project in Dunlop Estates Berhard. *Palm Growers Council 1st Technical Symposium*, Kuala Lumpur, Malaysia.

GREENSHIELDS, R.N. (1978). Review of progress of tower fermentation system for treatment of palm oil effluent and testing of by-product as potential animal feed. *Oil Palm Growers Council 3rd Technical Symposium*, Kuala Lumpur, Malaysia.

HALL, D.O. (1979). Solar energy conversion through biology. *Biologist* 26 (1) 16.

HALL, D.O. (1979). Solar energy conversion through biology. *Biologist* 26 (2) 67.

HESSELTINE, C.W. & WANG, H.L. (1967). Traditional fermented foods. *Biotechnology and Bioengineering*. 9, 275.

HESSELTINE, C.W. & WANG, H.L. (1979). Fermented Foods. *Chemistry and Industry* 12, 393.

HO, Y.C. (1978). Implementation of Regulations. *Oil Palm Growers Council 3rd Technical Symposium*. Kuala Lumpur, Malaysia.

KIRKCALDY, J. (1978). Oil recovery process for palm oil effluent treatment. *Oil Palm Growers Council 3rd Technical Symposium*, Kuala Lumpur, Malaysia.

LAWS OF MALAYSIA Act 127 (1974). *Environmental Quality Act*, Kuala Lumpur, Malaysia.

LAWS OF MALAYSIA (1977). *Environmental Quality (Prescribed Premises) (Crude Palm Oil) Regulations*. Tan Sri Ong Kee Hui Minister of Science, Technology and Environment, Kuala Lumpur, Malaysia.

NERANTZIS, E. & GREENSHIELDS, R.N. (1978). The continuous fermentation of vinegar by tower fermentation. *Ph.D. Thesis, The University of Aston in Birmingham*.

PETHIYAGODA, U. (1978). Coconut inflorescence sap. *Journal Perak Plantrs Association* 86.

SHEPHERD, R. (1977). Personal Communication.

STANTON, W.R. & WALBRIDGE, A. (1969). Fermented food processes. *Process Biochemistry* 4 (4) 45.

STANTON, W.R. (1972). Waste recovery by microorganisms. *UNESCO ICRO Work Study*. The University of Malaysia, Kuala Lumpur, Malaysia.

STANTON, W.R. (1978). Personal Communication. (British Patent pending).

WOOD, B.J.B. & MIN, Y.F. (1975). Oriental Food Fermentations. *The Filamentous Fungi*, Volume 1, *Industrial Mycology* (Eds. J. E. Smith and D.R. Berry, Arnold, London).

WOOD, B.J.B. (1978). Land disposal of palm oil mill effluents. *Oil Palm Growers Council 3rd Technical Symposium*, Kuala Lumpur, Malaysia.

BIOTECHNOLOGY IN THE SOVIET UNION AND SOME EASTERN EUROPEAN COUNTRIES

H. Michalski

Lodz Technical University, 90-924, Lodz, Poland

INTRODUCTION

Biotechnology is a very modern and new approach to a very old and traditional technology.

The development of microbial and biochemical processes has led to the necessity for the intensification and modernisation of such traditional processes. This, together with the introduction of new discoveries, has produced a need to solve technological problems created. Thus the extensive progress in technical biochemistry, applied microbiology and biochemical and microbial engineering has created a new area of science. As usual, even creation of a name for this new science was not easy to establish. In Russian technical literature many different terms were introduced such as Biotechnics, Bioengineering, Microbial Engineering and Biotechnology.

As an analogous term to Chemical Engineering, Microbial Technology or Biotechnology was introduced for all unit biochemical processes or for all technological processes used to obtain a final product. Similarly, all the required knowledge connected with unit operations and the equipment needed for these processes was called Bioengineering or Biotechnics.

THE SOVIET UNION

The first conference in the U.S.S.R. namely the All-Union Conference on Bioengineering and the All-Union Conference on Biotechnics was held in 1972. At these conferences it was decided that for the rapid development of biotechnological industries in the Soviet Union, great efforts in bioengineering research were necessary. At that time priority was given to tackling the following problems:

a. The discovery of the biological basis or principles involved in microbial synthesis, and the possibility of regulation of this process by the control of the external environment. In other words the production of equipment which would be specially designed to take into account the optimal statistics and dynamics of the system.

b. The investigation of the hydrodynamics, mass transfer - especially oxygen mass transfer, and the scaling-up of the fermenters.

c. The investigation of mathematical models for both microbial population growth and synthesis of useful metabolite together with the elaboration of the algorithms for process control by computers.

d. The development of the theory of both batch and continuous processes.

e. An investigation to develop or invent new equipment plus the actual construction of this, as a working apparatus for a typical technological process of microbial synthesis. The equipment to be first scrutinised would be sterilisation apparatus, air filters, fermenters, separators, evaporators, driers etc. Also to be considered was the setting up of the production lines for the microbial synthesis of products and to involve everything connected with the measurement, control and automation of such production lines.

Many Institutions and Research Centres throughout the Soviet Union were involved. The main theoretical and laboratory researches were carried out by the Institute of Biophysics and Physiology of the Microorganism of the Academy of Sciences of the U.S.S.R. Other work was undertaken by the other Institutes of the Academy of Sciences.

The practical aspects of such bioprocesses were investigated by the Industrial Research Centres such as the All-Union Research Institute for Antibiotics; the Moscow Institute of Chemical Equipment and Apparatus and the All-Union Institute of Protein Synthesis and many others.

As one particular example of the organisation and co-ordination of such research involved with the biotechnological problems in the Soviet Union can be shown in the Latvian Soviet Republic. The Latvian scientists were given the problem of the biosynthesis of biologically active substances and in particular amino acids such as L-lysine production. The organisation and co-ordination was carried out by a special Republican Co-ordination Committee. They directed what work was to be carried out by the Institute of Microbiology of the Latvian Academy of Sciences, the Biological Faculty of the National Latvian University, and the Latvian Academy of Agriculture and others.

During laboratory scale work all the problems which had previously been discussed as priorities were examined and solved. For example, from natural sources strains with a higher biosynthetic activity were isolated. Then after optimisation of the cultural conditions the activity was found to be greatly increased to 20 mg/ml of L-lysine, 25 mg/ml of L-alanine, 30 mg/ml of L-glutamic acid and 5 mg/ml of L-isoleucine and L-valine.

From this laboratory-scale work investigations went forward to pilot plant and larger scale levels with the ultimate aim of an experimental factory being built and equipped with most of the up to date fermentation systems. It was at this stage that the Latvian Academy of Sciences, together with the Institute of Biochemistry of the U.S.S.R., carried out a deep and detailed study of the microbial production of the feed concentrate, L-lysine.

Co-operation between these two Institutes led to a full technological process being worked out and a large pilot plant being built which yielded a productivity of 1,000 metric tonnes per year of L-lysine.

A special mutant strain of *Brevibacterium* was found with increased biosynthetic activity for L-lysine production. Optimisation of the medium content was only one of the many steps in the investigation. Further research showed great importance of the oxygen mass transfer intensity into the broth medium.

It was found that with both low and high aeration the same disadvantageous effect was seen. In both these cases a lowering of dehydrogenase activity was reported with the result that a diminished amount of L-lysine was produced. The maximum L-lysine production obtained from molasses broth was 66 g/l when only minimum by-products were produced by controlling the mixing and aeration according to the pH level of the culture medium.

The inhibitory effect of a large concentration of sucrose was eliminated by suitable feeding.

After taking into account all the above mentioned restrictions, maximum L-lysine production of 80 g/l under laboratory conditions was obtained. For the industrial scale special pneumatic fermenters of 100 m^3 capacity were designed, but great attention had to be paid to foam control.

Having considered the possibility of controlling the L-lysine production line by means of a computer, a mathematical model was evolved which took into account oxygen limitation and consisted of four differential equation systems. From this work it was possible to examine the influence of the advances in research on the progress of production technology and on the feedback effect science has on industrial results.

As it was a really outstanding research centre, the Institute of Microbiology of the Latvian Academy of Science in Riga was chosen to organise the All-Union Symposium on Biotechnology and Bioengineering in March 1978. This symposium was sponsored by the Academy of Sciences of the U.S.S.R., the Academy of Sciences of the Latvian Soviet Republic, the Centre for Management of Microbial Industry in the U.S.S.R., the Scientific Committee of Microbial Physiology and Biochemistry of the Academy of Sciences of the U.S.S.R. and the All-Union Microbial Society.

The main topics of the Symposium were: Fermentation processes; Media and products of microbial synthesis and Apparatus and equipment.

The significant advances in biotechnology and bioengineering in the Soviet Union can therefore be seen from the total of 346 abstracts. These were published in three separate volumes each volume covering one of the three main topics at the Symposium. Such a volume of work gives a picture of the outstanding effort of all scientific centres in the Soviet Union to cover this vast field of biotechnology and bioengineering and to help in the development of the microbial industry. In reading through these volumes of the Symposium one cannot fail to be amazed at both the number of scientific and research centres and the veritable army of scientists who are involved in these problems.

The topics covered in the three main main sections varied from the very theoretical to the very practical aspects of all the problems. For example, with amino acids production, and especially L-lysine production, the problems were still a subject of great interest. However, at that time the problems dealt with were the optimisation and improvement of the process based on using molasses as the substrate and on using a semicontinuous process.

In Moscow the All-Union Research Institute of Genetics and Selection for Industrial Microorganisms have studied the production of L-lysine on a semi-industrial scale. This was done as a batch process using acetic acid as the substrate. From this work a yield of 51 g/l was obtained. The product was prepared as a concentrate containing 72% lysine monochlorohydrate. Such a product, after an ion-exchange treatment, gave a crystalline form containing the equivalent of 97% lysine.

However, it is not only the biochemical, microbiological or biological centres which are involved. Also caught up in these advancing programmes are such Institutions as the

Research Institute for Industrial Automation at Grozngi, the Kiev Institute of Automation, the Woronez Institute of Technology, the Research Institute of Chemical Equipment at Irkutsk, the Krasnojarsk Institute of Physics of the U.S.S.R. Academy of Sciences and the Chemical-Technological Institute at Kazansk. These are spread over the whole of the Soviet Union.

To give a complete picture of all the topics covered at the various sessions would involve too much time. However, some of the main topics of the lectures are listed below.

The limits of the capability of microorganisms in the synthesis of SCP and their attainability by technical means.

Continuous cultivation of microorganisms.

Thermodynamics of fermentation processes.

The role of oxygen in the process of microbial growth.

Rheological problems in deep fermentations involving filamentous fungi.

The mixing, mass transfer and scale-up problems in biosynthetic processes.

The regulatory effect of carbon dioxide on microbial cultivation.

The problem of foam in a fermentation process.

The continuous cultivation of microorganisms together with growth inhibition.

The kinetics of the biosynthesis of microbial depolymerases on media which do not contain suitable substrates.

The production of L-lysine on an acetic acid medium.

The preparation of media consisting of ligno-cellulose for SCP production.

Glucose-isomerase and the prospects for its application.

Extraction and purification of metabolites.

Disintegration of microorganisms.

Anaerobiosis as a method of microbial preservation.

The use of computers in the control of microbial plants.

Selective electrodes.

Methods for optimising microbial fermentations.

The apparatus and equipment for bakers yeast production.

The equipment for fodder yeast production on media with hydrolysates.

The theory and design of drying apparatus for the products of microbial synthesis.

Economic and technological aspects of organic acid biosynthesis.

From the above can be deduced that a vast amount of time, effort and money is being invested in Biotechnology and Bioengineering in the Soviet Union.

POLAND

As in other Eastern European countries, research in Poland is split among the different Institutions. Basic and more theoretical work is carried out in the laboratories of the Polish Academy of Sciences and in different Departments of the Universities. The majority of applied research and some theoretical work is the province of the Technical Universities, because they usually have a direct link with industry. Industrial Research Institutes and Research Centres are mainly involved in current industrial production problems.

According to their importance all the research problems are divided into three categories:

Category 1 or state problems which are very general and basic. These can be further subdivided into sections which are more detailed. For example - the state or first category problem could be Protein Production with every research effort to increase the quantity of protein coming under this particular heading. At the sublevels the topics undertaken are, for instance, L-lysine production based on molasses from Polish industry or from unconventional substrates, SCP from methanol etc. These latter topics are undertaken by a variety of different research centres which have different specialisations such as analytical chemistry, electrical engineering, biochemistry, mechanical engineering etc.

Category 2 problems are on the level of a particular Ministry.

Category 3 are factory problems. Such problems of interest may involve a particular factory or similar factories. For example, such a problem could be the storage of molasses from beet sugar factories.

As with the Soviet Union, many Scientific and Research Institutes in Poland are involved in the progress of biotechnology. For example, at the Industrial Research Institute of Fermentation Technology in Warsaw a great deal of work is connected with the current production of alcoholic beverages such as beer and also with bakers yeast, fodder yeast, citric acid etc. This Institute acts as the main co-ordinator for research connected with SCP production from both natural and unconventional substrates such as methanol etc.

In this co-ordination capacity let me give three examples which have involved other Institutions.

1. SCP production from some waste products of forestry and the paper industry has been a subject of a large amount of research at the Research Institute of the Paper Industry.

2. Using a selected and improved strain of *Aspergillus niger* a deep fermentation system for citric acid production has been examined on a large pilot plant scale. This work involved a number of Departments from the Technical University of Lodz; the Economic and Engineering Academy at Wroclav and the Industrial Research Centre for Citric Acid Production in Zgierz.

3. Many improvements and new discoveries were made by the Industrial Research Institute of Pharmacy along with the Industrial Institute for Industrial Chemistry in Warsaw and various Departments from different Universities throughout Poland. Such work has helped and improved the production of certain pharmaceuticals and antibiotics.

The Institute for Technical Biochemistry at the Technical University of Lodz has organised the XVIth Conference of the Polish Biochemical Society in September 1978.

Many topics were discussed and included:

The investigations on bacterial α-amylase production on a laboratory, pilot plant and semi-industrial scale and the applications of this bacterial enzyme in the textile industry.

The biosynthesis of β-glucanase by *Trichoderma viride*.

The application of statistical methods for the measurement of volumetric oxygen mass transfer coefficient (k_La) in fermentation processes.

The immobilisation of α-amylase and rennet on polyacrylamide gel by a radiation method.

CZECHOSLOVAKIA

Czechoslovakia is a country with a very long tradition in biotechnological studies. Not only is its pilzner beer world famous, but the developed centre for continuous culture for microorganisms is also well known. Many Czech scientists are world famous because of their connection with continuous culture work. Such workers introduced the theoretical and practical uses of chemostats and turbidostats and also the theoretical aspects of multistage and multistream cultures and fermentations. Their approach to such work has provided a very useful tool for the elaboration and prediction of the results from very complicated fermentations involving many antibiotics, secondary metabolites, etc.

Academician Malek, the father of continuous culture, is a member of the Czechoslovak Academy of Sciences and founder of the Institute of Microbiology in Suburb Krc in Prague where much outstanding work has been done. One branch of the Institute is the Department of Applied Microbiology where scientists such as Sobotka, Prokop, Votruba, Novak, Sykita, Ricica and others work. Such people are continually submitting extremely interesting papers to the various biotechnological journals and have dealt with oxygen mass transfer problems, methods for measuring dissolved oxygen, mathematical models and the influence of different phenomena on readings of dissolved oxygen probes.

At the Prague Institute of Chemical Technology in the Department of Fermentation Chemistry there is a very strong Biotechnology Unit. At the XIIth International Congress of Microbiology in Munich an extremely interesting paper was presented by this group in connection with SCP production. Using different yeast strains experiments were carried out in both synthetic media and sulphite waste liquor media. Such experiments were carried out on a continuous laboratory and pilot-plant scale. By applying relatively low concentrations of ethanol it became possible to improve significantly the economics of this process.

The antibiotic industry is another of the major interests involving biotechnology in CSSR. It was for this reason that a Research Institute for Antibiotics and Biotransformations was set up at Roztoky near Prague.

HUNGARY

In Hungary, a leading centre of biotechnological research is to be found in the Institute of Agricultural Chemical Technology which is part of the University of Technical Sciences in Budapest. Their main interest is the optimisation of SCP production from methanol using a computer controlled fermentation system. The efficiency of such a system is presented by static optimisation experiments using continuous fermentation based on the turbidostat principle producing SCP from a single carbon source.

The Biogal Pharmaceutical Works - a Division of the Research and Technical Progress in Debrecen are also interested in the introduction of computer controlled fermenter systems but on a bigger scale. They have been developing an off-line system consisting of a 1.5 m^3 capacity fermenter linked to a high speed data collection system and a desk calculator for the logging and evaluation of the process data. Auto-analysers have been installed in order to increase the number of biochemical parameters covered.

At the Institute of Haematology and Blood Transfusion in Budapest a large scale plant has been developed for the fractionation of plasma protein.

Again some research on continuous culture has been carried out by Dr Zetchelaki at the Food Research Institute in Budapest. This work was presented at the IVth FEMS Symposium in Vienna in 1977.

GERMAN DEMOCRATIC REPUBLIC

As Czechoslovakia has Academician Malek, so the German Democratic Republic has Professor F. Bergter at the Central Institute of Microbiology and Experimental Therapy in Jena which is part of the G.D.R. Academy of Sciences. He is a specialist in the kinetics of microbial growth and his fundamental book called *The Growth of Microorganisms* is invariably cited in text books on Biochemical Engineering. His kinetic model of mycelial growth was presented at the VIth FEMS Symposium in Vienna.

Other workers in G.D.R. involved in the Biotechnology field are the group at the Institute of Technical Chemistry in Leipzig. The head of this group is Professor M. Ringpfail who gave papers at the Biotechnology and Bioengineering Conference in Riga in 1978. Other members of the group, Babel and others, presented papers on the regulation and dissimilatory sequences of methylotrophic bacteria at the IInd International Symposium on Microbial Growth on Single Carbon Compounds held in Pushchino near Moscow in 1977.

Professor Ringpfail's group are mainly interested in SCP production and its technical problems; automatic control of processes involved in cell mass synthesis; and methods of optimisation of microbial fermentations. Another group, from the Institute of Technical Chemistry of the G.D.R. Academy of Sciences, also presented papers at the II nd International Symposium. These papers dealt with a thermodynamic description of fermentation processes. It was at this Institute of Technical Chemistry that the JZ deep jet aeration system was invented and the industrial high performance fermenters for aerobic fermentations were developed.

YUGOSLAVIA

Yugoslavia is also heavily involved in Biotechnology and Bioengineering research. This is mainly because of major interests in the Pharmaceutical Industry. The inspiration and vital enthusiasm for research in this field is demonstrated by the group at the Faculty of Technology at Zagreb University. Heading this group is Professor Vera Johanides, perhaps comparable to Malek and Bergter. She has been involved in educating and inspiring many of the outstanding biotechnology scientists in this country. In collaboration with Mervic and Maric many strains have been isolated from natural sources for SCP. One strain, namely *Candida lipolytica* 33M, grows well on methanol.

This work also examined the influence of formaldehyde and formic acid growth and also the specific oxygen uptake. This group, together with the INA Research and Development, have been investigating the activity of copper and oxygen on the obligate methylotropic bacterial strain 4025 grown on methanol.

In the Faculty of Agriculture at Belgrade University the application of mixed cultures as a method of increasing the yield of penicillin and streptomycin is being studied. Cimerman and others from the Kemijaki Institute of Boriz Kidnc in Ljubljana have investigated citric acid fermentation and mycelial differentiation in submerged citric acid production.

At the Department of Chemistry at Ljubjana University the relationship between growth and acid production of acetic acid bacteria has been studied. It was found that acetic acid was used up and was not produced as a result of growth metabolites.

PLIFA - The Pharmaceutical and Chemical works at Zagreb - have worked on research involving the growth kinetics of *Aspergillus niger* and glucoamylase biosynthesis in repeated batch culture. They have also examined the hydrolysis of some cellulosic materials using the culture filtrates of a fungus isolated from soil.

In the Faculty of Technology of Zagreb University they have examined the growth of higher fungi on whey in submerged culture. Using a diluted whey with 50% water they were able to obtain very good growth of *Agaricus campestris* and *Morchella hortensis* growing about 7-8 g of mycelial dry weight per litre.

SPECIES INDEX

SUBJECT INDEX